Occupational Exposures

Occupational Exposures: Chemical Carcinogens and Mutagens offers a focused emphasis on chemical exposures associated with carcinogenic and mutagenic impacts along with associated controls for mitigating and controlling exposures. It discusses a range of topics including hematopoietic system impact, reproductive system impact, inorganic compounds, halogenated compounds, carbamates, polycyclic aromatic hydrocarbons, aromatic amines, product elimination and substation, exposure control methods, and human biological impact.

- Presents a comprehensive account of carcinogens and mutagens for occupational and environmental health professionals
- Covers preventive measures and controls for carcinogens and mutagens
- Discusses exposure controls, exposure pathways, impacts, and treatments

The book is ideal for professionals and graduate students in the fields of occupational health and safety, industrial engineering, and chemical engineering.

Sustainable Improvements in Environment Safety and Health

Series Editor:
Frances Alston
California, USA

Lean Implementation
Applications and Hidden Costs
Frances Alston

Safety Culture and High-Risk Environments
A Leadership Perspective
Cindy L. Caldwell

Industrial Hygiene
Improving Worker Health through an Operational Risk Approach
Frances Alston, Emily J. Millikin, and Willie Piispanen

The Legal Aspects of Industrial Hygiene and Safety
Kurt W. Dreger

Strategic Environmental Performance
Obtaining and Sustaining Compliance
Frances Alston and Brian Perkins

Occupational Exposures
Chemical Carcinogens and Mutagens
Frances Alston and Onwuka Okorie

For more information about this series, please visit: https://www.crcpress.com/
Sustainable-Improvements-in-Environment-Safety-and-Health/book-series/
CRCSUSIMPENVSAF

Occupational Exposures
Chemical Carcinogens and Mutagens

Frances Alston and Onwuka Okorie

CRC Press
Taylor & Francis Group
Boca Raton London New York

CRC Press is an imprint of the
Taylor & Francis Group, an **informa** business

First edition published 2023
by CRC Press
6000 Broken Sound Parkway NW, Suite 300, Boca Raton, FL 33487-2742

and by CRC Press
4 Park Square, Milton Park, Abingdon, Oxon, OX14 4RN

CRC Press is an imprint of Taylor & Francis Group, LLC

Library of Congress Cataloging-in-Publication Data
Names: Alston, Frances (Industrial engineer), author. | Okorie, Onwuka, author.
Title: Occupational exposures : chemical carcinogens and mutagens / Frances Alston and Onwuka Okorie.
Description: First edition. | Boca Raton : CRC Press, 2023. |
Series: Sustainable improvements in environment safety & health |
Includes bibliographical references and index.
Identifiers: LCCN 2022036795 | ISBN 9781032054506 (hardback) |
ISBN 9781032114910 (paperback) | ISBN 9781003220114 (ebook)
Subjects: LCSH: Industrial safety. | Chemical industry–Health aspects. |
Chemical industry–Safety measures.
Classification: LCC RC965.C44 A425 2023 | DDC 363.11–dc23/eng/20221021
LC record available at https://lccn.loc.gov/2022036795

ISBN: 978-1-032-05450-6 (hbk)
ISBN: 978-1-032-11491-0 (pbk)
ISBN: 978-1-003-22011-4 (ebk)

DOI: 10.1201/9781003220114

Typeset in Sabon
by codeMantra

Contents

Authors

Dr. Frances Alston is an Adjutant Professor at Oregon State University and the 2018 President of the American Society for Engineering Manager. She is also a Fellow of the American Society for Engineering Management (ASEM) and has built a solid career leading the development and management of Environment, Safety, Health, and Quality (ESH&Q) programs in diverse cultural environments.

Dr. Onwuka Okorie is an Industrial Hygiene Professional and Subject Matter Expert at the Lawrence Livermore National Laboratory (LLNL) in Livermore, California. He is a Certified Industrial Hygienist and a Certified Safety Professional, and a Fellow of the Royal Society for Public Health (RSPH).

Chapter 1

Polycyclic aromatic hydrocarbons

1.1 INTRODUCTION

Polycyclic aromatic hydrocarbons (PAHs) are compounds containing only carbon and hydrogen atoms. PAHs are comprised of two or more benzene rings chemically bonded in a linear, cluster, or angular arrangement.[1,2] PAHs is an expression used to denote a group of chemically related organic compounds of various structures with varying toxicities. PAHs typically exist as colorless, white, or pale-yellow solids and are produced when substances are burned. They are formed from incomplete combustion of organic substances such as coal tar, gas, or crude oil. More PAHs are given off when the burning process is less efficient. For example, when there are forest fires and volcanoes, PAHs are produced naturally.[7,10] These substances may also be manufactured and used to make plastics or dyes. PAHs also exist in products made from fossil fuels, such as coal, natural gas, and asphalt. When coal is converted into natural gas, PAHs are released. There may be high levels of PAHs in some coal-gasification sites. PAHs also can be released into the air during the burning of fossil fuels, garbage, or other organic substances. The inherent properties of PAHs such as heterocyclic aromatic ring structures, hydrophobicity, and thermostability have made them recalcitrant and highly persistent in the environment. PAH pollutants have been determined to be highly toxic, mutagenic, carcinogenic, teratogenic, and immunotoxicogenic to humans.[11,12,14]

PAHs are a class of chemicals composed of up to six benzene rings fused together such that any two adjacent benzene rings share two carbon bonds as shown in Figure 1.1.

Individual PAH compounds do not have the exact same health effects. Table 1.1 shows some existing information about where these PAHs are in existence and any known health effects.[7]

Cooking of meat and other foods can produce PAHs. Naphthalene is a PAH that is produced commercially in the United States to make other chemicals. Cigarette smoke contains many PAHs that have toxic, mutagenic, and carcinogenic properties. PAHs are highly lipid soluble and thus readily absorbed from the gastrointestinal tract. The physical features of

DOI: 10.1201/9781003220114-1

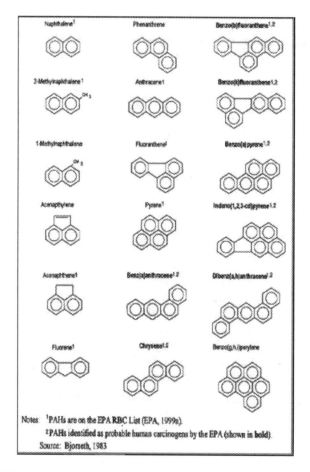

Figure 1.1 PAHs.

PAHs are high melting and high boiling points, low vapor pressure, and very low aqueous solubility. The vapor pressure and aqueous solubility tend to decrease with increasing molecular weight. Aqueous solubility of PAHs decreases for each additional ring. PAHs are very soluble in organic solvents because they are highly lipophilic. These substances also manifest various functions such as light sensitivity, heat resistance, conductivity, emit ability, and corrosion resistance. Depending on their molecular weight, they are emitted either as gaseous-phase low molecular weight (LMW) PAHs or in the particulate form as high molecular weight (HMW) PAHs. Table 1.2 lists general usage for some PAHs.

There are three classes of PAHs in the environment namely biogenic PAHs that are produced by living organisms, like vegetation; petrogenic

Table 1.1 PAH sources

PAH	Cigarettes/ e-cigarettes	Diesel fuels	Grilled/ charred foods	Petroleum products, coal tar, and coal burning	Wildlife/ agricultural smoke	Health concerns
Acenaphthene	☒	☒	☒	☒	☒	Irritant
Acenaphthylene	☒	☒	☒	☒	☒	No information available
Anthracene	☒	☒	☒	☒	☒	No information available
Benz[a]anthracene	☒	☒	☒	☒	☒	Cancer risk
Benzo[a]pyrene	☒	☒	☒	☒	☒	Cancer risk
Benzo[b]fluoranthene	☒	☒	☐	☒	☒	Cancer risk
Benzo[e]pyrene	☒	☐	☒	☒	☒	No information available
Benzo[ghi]perylene	☒	☐	☒	☒	☒	No information available
Benzo[j]fluoranthene	☒	☒	☒	☒	☒	Cancer risk
Benzo[k]fluoranthene	☒	☒	☒	☒	☒	Cancer risk
Chrysene	☒	☒	☒	☒	☒	No information available
Dibenzo[a,h]anthracene	☒	☒	☒	☒	☒	Lung irritant
Fluoranthene	☒	☒	☒	☒	☒	Cancer risk
Fluorene	☒	☒	☒	☒	☒	Irritant
Naphthalene	☒	☒	☒	☒	☒	Irritant
Phenanthrene	☒	☒	☒	☒	☒	Irritant
Pyrene	☒	☒	☒	☒	☒	Irritant
Indeno[1,2,3-cd]pyrene	☒	☒	☒	☒	☒	Cancer risk

Potential sources of PAHs

Table 1.2 PAH general usage

PAH	General usage
Acenaphthene	Manufacture of pigments, dyes, plastics, pesticides, and pharmaceuticals
Fluoranthene	Manufacture of agrochemicals, dyes, and pharmaceuticals
Fluorene	Manufacture of pharmaceuticals, pigments, dyes, pesticides, and thermoset plastic
Phenanthrene	Manufacture of resins and pesticides
Pyrene	Pyrene

PAHs that are generated by geological processes that could be natural as in seeps, coal outcrops, or anthropogenic as in crude oil spills, fossil fuel release; and pyrogenic PAHs that are generated by high temperature combustion of organic matter occurring naturally as forest fires, prairie fires or anthropogenic as wood stoves, car exhaust, and coal tar.

1.1.1 Benzene

Benzene is an aromatic hydrocarbon that is widely used as a solvent in chemical laboratories and as an intermediate in the chemical industry for manufacturing polymers and other products. The CAS number for benzene is 71-43-2 having the chemical formula of C_6H_6. Benzene is a common contaminant in the atmosphere as approximately 99% of the benzene present in the human body occurred through inhalation.[16.] The structure of benzene and the benzene ring is shown in Figure 1.2.

Benzene, also commonly known as benzol, is a colorless liquid with a sweet pungent odor. Benzene is found in air, water, and soil that comes from both industrial and natural sources, is highly flammable, and evaporates into air very quickly, and dissolves slightly in water.

Most people can begin to smell benzene at around 60 parts per million (ppm) in air and taste benzene in water at 0.5–4.5 ppm. Various industries use benzene to make other chemicals, such as styrene, cumene, and cyclohexane. In addition, benzene is also used in the manufacturing of some rubbers, lubricants, dyes, detergents, drugs, and pesticides. Natural sources of benzene include gas emissions from volcanoes and forest fires. These natural sources contribute to the presence of benzene in the environment. Benzene is also present in crude oil, gasoline, and cigarette smoke.

Benzene can enter the body through lungs via inhalation, the gastrointestinal tract, and the skin. When exposure to high levels of benzene in air occurs, about half of the benzene breathe in passes through the lining of the lungs and enters the bloodstream. When food or drink is exposed to benzene, most of the benzene taken in by mouth passes through the lining of the gastrointestinal tract and then enters the bloodstream. A small

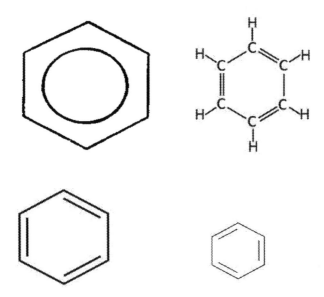

Figure 1.2 Benzene structures.

amount of benzene will enter the body by passing through the skin and into the bloodstream during skin contact with benzene and benzene-containing products. Once benzene is in the bloodstream, benzene travels throughout the body and can be temporarily stored in the bone marrow and fat.

After exposure to benzene, there are primary two factors that can determine if harmful health effects will occur and the severity of effects. These factors include the amount of benzene to which an individual is exposed and the length of time of the exposure.

Brief exposure between 5 and 10 minutes to extremely high levels of benzene in air (10,000–20,000 ppm) can result in death, whereas lower levels between 700 and 3,000 ppm can cause drowsiness, dizziness, rapid heart rate, headaches, tremors, confusion, and unconsciousness.

Eating foods or consuming liquids containing high levels of benzene can cause vomiting, irritation of the stomach, dizziness, sleepiness, convulsions, rapid heart rate, coma, and even death. When benzene is spilled on the skin, it can cause redness and sores. If benzene comes in contact with the eyes, it may cause general irritation and damage to the cornea. The Department of Health and Human Services (DHHS), the International Agency for Cancer Research (IACR), and the Environmental Protection Agency (EPA) have determined that benzene is carcinogenic to humans. Children can be affected by benzene exposure in the same ways as adults. Benzene can pass from the mother's blood to a fetus.[15]

1.2 EXPOSURE PATHWAYS

The main routes of exposure for PAHs include ingestion, inhalation, and dermal exposure that could be work-related or occur in a non-occupational environment. Occupational exposure may also occur in workers who breathe exhaust fumes, such as motor vehicle drivers or mechanics, and workers who work in mining or refining industries. Some worker exposures may involve more than one route at the same time such as dermal and inhalational exposures leading to an increase in the total absorbed dose.[1] Workers can be exposed to PAHs in the air and surface soil through inhalation, ingestion, or dermal contact.

The main means of exposure to PAHs in the general population is from breathing ambient and indoor air, smoking cigarettes, or ingestion of food containing PAHs. Smoking is an important route of exposure as tobacco smoke contains different PAHs, such as benzo(a)pyrene, naphthalene, fluorene, phenanthrene, and many other known or suspected human carcinogens. Some plants such as wheat, rye, and lentils can synthesize PAHs or absorb them from water, air, or soil. Water can also contain certain amounts of PAHs as these chemicals can get into the water from the soil or from industrial sewage and accidental marine pollution during oil shipping. The soil can contain PAHs, mainly from airborne effect. As a result, most people are regularly exposed to PAHs. People should avoid smoke from wood fires, avoid exposure to automobile exhaust and areas of high traffic congestion, and avoid areas where asphalt road construction or tar roofing is occurring. Smoking should be done outdoors and in areas away from children. Foods should be prepared by slower cooking over low heat, rather than by charring or grilling food and kin contact with soot should be avoided.

There are several sources of data for PAH exposure surveillance that are maintained by the Occupational Safety and Health Administration (OSHA) such as the Integrated Management Information System (IMIS) and the Occupational Exposure Databanks (OEDBs). An exposure evaluation worksheet is included in Appendix A to assist in determining potential exposure routes.

1.2.1 Inhalation exposures and impact

The main sources of exposure to PAHs for most of the US population are the inhalation of the compounds from tobacco smoke, other smoke sources like wood smoke, ambient air, and the consumption of PAHs in foods. Workers can be exposed to PAHs by inhaling engine exhaust gas and using products that contain PAHs in a variety of industries such as mining, refining, chemical, transportation, and electrical engineering. PAHs have also been found in places where petroleum, petroleum products, and coal are used,[3] which also presents inhalation hazard potential.

1.2.2 Dermal exposures and impact

The occupational exposure assessment is mainly based on the air concentrations of the chemicals. However, some chemicals also have a skin notation in the occupational exposure limit (OEL) values, which indicates that these chemicals can enter the body through the skin. Skin contamination of PAH occurs in road paving workers working with petroleum asphalt. Worker skin contamination appeared to increase ten-fold if the asphalt contained coal tar.[5] In some work environments where PAH exposure has been high, skin contamination by PAHs has demonstrated a stronger correlation with urinary 1-hydroxypyrene (1-OHP) excretion than with the PAH concentration through inhalation.

1.2.3 Ingestion exposure and impact

Food ingestion is the main route of exposure compared with inhalation for a large proportion of general population exposed to PAHs. In addition to occupational PAH exposure, diet makes a significant contribution of PAH intake in non-smokers. The environment and food processing techniques are sources of PAH contamination of food. The accumulation of PAHs in foods of animal origin, especially livestock, is due to the consumption of contaminated pastures and vegetation. The types of food consumed, composition of gastrointestinal fluids, and transport processes across intestine influence the bioavailability of PAHs. The metabolites of PAHs with two and three rings are excreted mainly in the urine, whereas higher-molecular-weight metabolites are excreted in the feces. Liver is the main organ for PAH metabolism. However, depending on the route of entry of PAH, other organs may play a more important role. If ingested, the intestinal flora and intestinal cytochrome P450 enzymes can promote the metabolism of PAHs.[6]

1.3 CHEMICAL TO HUMAN INTERFACE

As a result of the high lipophilicity of PAH, they have a significant bioavailability after ingestion and inhalation. Studies have shown that PAH levels have been found in most internal organs, especially organs that are rich in adipose tissue. These organs can serve as reservoirs from which the hydrocarbons can gradually be released. Once they enter the body, PAHs require a multistep metabolic activation by specific enzymes. The main enzyme system responsible for PAH metabolism is the mixed-function oxidase system. The first reaction is epoxidation. PAH epoxides can then be conjugated with glutathione, and this is considered a true detoxification reaction. The epoxides that are not conjugated with glutathione are converted into phenols and diols. These PAH metabolites are sometimes not polar enough to be excreted and, therefore, must be conjugated with glucuronic or sulfuric acids to enable excretion. Most metabolites of PAH are excreted in feces and urine.[9]

1.4 EXPOSURE CONTROL METHODS

Controlling occupational exposure to PAH which is considered a carcinogen is a principal step in decreasing occupational cancer burden. The Hierarchy of Controls has been used as an exposure control strategy. Exposure control strategies are ranked from most effective (elimination or substitution/engineering) to least effective (personal protective equipment [PPE]). An effective control strategy uses multiple controls from the Hierarchy and often includes monitoring to evaluate the effectiveness of the control program. The OSHA regulated exposures to PAHs under OSHA's Air Contaminants Standard for substances termed coal tar pitch volatiles (CTPVs) and coke oven emissions. Employees exposed to CTPVs in the coke oven industry are covered by the coke oven emissions standard. The OSHA coke oven emissions standard requires employers to control employee exposure to coke oven emissions using engineering controls and work practices. Wherever the engineering and work practice controls, which can be instituted are not sufficient to reduce employee exposures to or below the permissible exposure limit, the employer shall nonetheless use them to reduce exposures to the lowest level achievable by these controls and shall supplement them using respiratory protection. The OSHA standard also includes elements of medical surveillance for workers exposed to coke oven emissions. The OSHA permissible exposure levels (PELs) for PAHs in the workplace is 0.2 mg/m^3 for 8-hour time-weighted average (TWA). The National Institute for Occupational Safety and Health (NIOSH) has recommended that the workplace exposure limit for PAHs be set at the lowest detectable concentration, which was 0.1 mg/m^3 (REL=recommended exposure limit) for CTPV agents for a 10-hour workday, 40-hour workweek.[9] One of the main sources of exposure to PAHs is the inhalation of these compounds through tobacco smoke. Smokers can lower their own exposure and the exposure of their families by quitting smoking. People could also reduce their use of wood stoves and fireplaces. Other steps to lower exposure to PAHs include the following:[7]

- Decrease consumption of smoked foods or foods cooked with charcoal.
- Decrease use of coal-tar-based cosmetics and shampoos.
- Use of cedar shavings or aromatic herbs instead of mothballs, moth flakes, and deodorant cakes.
- Wear protective clothing and gloves if you are handling creosote-treated wood products.
- Wear an appropriate respirator when working with products containing PAHs.

1.4.1 Product elimination and substation

Biomonitoring plays an important role, but the complexity of PAH mixtures, their low concentration in biological media, and their metabolism make biomonitoring of the components of the mixture difficult. The biological monitoring methods of PAH exposure include the measurement of DNA, hemoglobin, other PAH adducts in urine, serum protein, and PAH metabolites in urine. Several different remediation technologies have been tested in efforts to remove PAHs from the environment. One of such technologies is bioremediation, which is safe. A selection of genera of gram-positive and gram-negative bacteria, fungi, and algae have been isolated and characterized for their ability to use PAHs.

1.4.2 Engineering controls

Engineering controls are the best way to control exposure to PAHs. Engineering controls include local and dilution ventilation, isolation and/or containment processes that emit PAHs, and where possible, automated handling of coal tar products that generate CTPVs in open systems. Controls also include ensuring workers are enclosed and separated from atmosphere contaminated with PAHs such as enclosed rooms that are pressurized and supplied with filtered air.

1.4.3 Administrative controls

Administrative controls must be considered if engineering controls are not feasible or do not provide adequate control of exposures. Such controls include worker training in safety procedures to prevent PAH exposures, adjusting work tasks or schedules that involve exposure to PAHs, written operating procedures, work practices, and safety and health rules that workers must follow to complete jobs involving PAHs safely.

1.4.4 Protective equipment and effectiveness

The use of PPE to protect workers from exposure to PAH is a vital component of an effective exposure control program. PPE is often used because it is readily accessible and economical even though it is the least effective and the lowest in the hierarchy of exposure controls to assure workers are protected against workplace hazards associated with PAH. Respirators may also be worn by individuals exposed to CTPVs to keep their exposures below the OSHA PEL, and protective clothing can be worn to prevent skin contact with coal tars and coal tar products, including CTPVs. However, PPE should be the last line of defense. Respiratory protective equipment (RPE) is used to control inhaled PAH exposure. However, when PAHs are absorbed through the skin, exposure may occur. Evaluating only airborne exposure may not reflect the complete exposure situation. All exposure pathways must be taken into consideration to establish a complete exposure profile.

1.5 HUMAN BIOLOGICAL IMPACT

PAHs are regarded as a human health issue. Occupational exposures to high levels of pollutant mixtures containing PAHs have resulted in symptoms such as eye irritation, nausea, vomiting, diarrhea, and confusion. Mixtures of PAHs are known to cause skin effects in animals and humans such as irritation and inflammation. Several studies show increased incidence of lung, skin, and urinary cancers in humans that have been exposed to PAH mixtures. There is a close correlation between human lung cancer and inhaled PAHs.[4] Many workers occupationally exposed to PAHs may have caused various health problems including cancers as many PAH compounds have been classified as probable or possible carcinogens by regulatory bodies such as the National Toxicology Program and the EPA.

Many studies on PAH have found potential links to certain cancers but have not explicitly proven that these agents are causative, as there is insufficient data on individual PAHs and mixtures. Most PAHs are not only pro-carcinogens but also are listed as genotoxic and mutagenic. The carcinogenicity of PAHs, as well as their toxic equivalency factor (TEF), is related to one of the most significant pyrogenic carcinogens, benzo[a] pyrene (B[a]P).

As PAHs have been implicated as causative agents of breast, lung, and colon cancers and have been associated with neuro-, reproductive, and developmental toxicities, the processes governing the disposition of these chemicals in the body and their subsequent metabolic fate assume a greater importance.[6]

Exposure to PAHs may pose a risk not only for lung cancer but also for cardiovascular diseases, including atherosclerosis, hypertension, thrombosis, and myocardial infarction.[8]

Long-term exposure to low levels of some PAHs has caused cancer in laboratory animals. Benzo(a)pyrene is the most common PAH to cause cancer in animals. Studies of workers exposed to mixtures of PAHs and other compounds have noted an increased risk of skin, lung, bladder, and gastrointestinal cancers. The information provided by these studies is limited because the workers were exposed to other potential cancer-causing chemicals besides PAHs. Although animal studies have shown adverse reproductive and developmental effects from PAH exposure, these effects have generally not been seen in humans.[7]

1.5.1 Potential impact to exposed individual

PAHs affect organisms through various toxic actions. The mechanism of toxicity is through the interference with the function of cellular membranes as well as with enzyme systems which are associated with the membrane. They cause carcinogenic and mutagenic effects and are potent immunosuppressants. Effects have been documented on immune system development, humoral immunity, and host resistance. The most extensively studied PAHs are 7,12-dimethylbenzo anthracene (DMBA) and benzo(a)pyrene (BaP).[9]

Although unmetabolized PAHs can have toxic effects, a major concern is the ability of the reactive metabolites, such as epoxides and dihydrodiols, of some PAHs to bind to cellular proteins and DNA. The resulting biochemical disruptions and cell damage lead to mutations, developmental malformations, tumors, and cancer. Evidence indicates that mixtures of PAHs are carcinogenic to humans. The evidence comes primarily from occupational studies of workers exposed to mixtures containing PAHs, and these long-term studies have shown an increased risk of predominantly skin and lung as well as bladder and gastrointestinal cancers. However, it is not clear from these studies whether exposure to PAHs was the main cause as workers were simultaneously exposed to other cancer-causing agents. Benzo(a)pyrene is the most common PAH to cause cancer in animals, and this compound is notable for being the first chemical carcinogen to be discovered. Based on the available evidence, both the International Agency for Research on Cancer (IARC, 1987) and US EPA (1994) classified a few PAHs as carcinogenic to animals and some PAH-rich mixtures as carcinogenic to humans. The EPA has classified seven PAH compounds as probable human carcinogens: benz(a)anthracene, benzo(a)pyrene, benzo(b)fluoranthene, benzo(k) fluoranthene, chrysene, dibenz(ah)anthracene, and indeno(1,2,3-cd) pyrene.[9] Genotoxic effects for some PAH have been demonstrated both in rodents and in vitro tests using mammalian (including human) cell lines. Most of the PAHs are not genotoxic by themselves, and they need to be metabolized to the diol epoxides, which react with DNA, thus inducing genotoxic damage. Genotoxicity plays an important role in the carcinogenicity process and maybe in some forms of developmental toxicity as well.[9]

1.5.2 Potential impact to offsprings

Embryotoxic effects of PAHs have been described in experimental animals exposed to PAH such as benzo(a)anthracene, benzo(a)pyrene, and naphthalene. Laboratory studies conducted on mice have demonstrated that ingestion of high levels of benzo(a)pyrene during pregnancy resulted in birth defects and decreased body weight in the offspring. It is not known whether these effects can occur in humans. However, the Center for Children's Environmental Health reports studies that demonstrate that exposure to PAH pollution during pregnancy is related to adverse birth outcomes including low birth weight, premature delivery, and heart malformations. High prenatal exposure to PAH is also associated with lower IQ at age 3, increased behavioral problems at ages 6 and 8, and childhood asthma. Cord blood of exposed babies shows DNA damage that has been linked to cancer.[9]

1.6 TREATMENT OF RESULTANT ILLNESS

The treatment of individuals exposed to PAHs differs for acute and chronic dose exposures. The primary goals after acute high-dose PAH exposure are decontamination and supportive measures. Treatment of chronic PAH toxicity is symptomatic and supportive. Health education and risk communication are important aspects of patient care. Contaminated clothing should be removed from the victim as soon as possible. The victim's skin should be decontaminated by gently scrubbing with soap and water. Ocular contamination should be treated with irrigation and a complete eye examination. Supportive care should be administered as clinically necessary.

Individuals exposed to high levels of PAHs have an increased risk for bronchogenic cancer. Periodic evaluations of healthy individuals exposed to PAHs may help with the early diagnosis and management of malignancy. Medical surveillance is recommended for workers exposed to PAH. The management of PAH-related illness can be done by occupational health clinicians who can recognize, evaluate, and treat illnesses resulting from exposure to PAHs.[13]

REFERENCES

1. Arey, J., & Atkinson, R. (2003). Photochemical reactions of PAHs in the atmosphere. *PAHs: An Ecotoxicological Perspective*, 47–63. New York: John Wiley & Sons, Inc.
2. Di Toro, D.M., McGrath, J.A., & Hansen, D.J. (2000). Technical basis for narcotic chemicals and polycyclic aromatic hydrocarbon criteria. I. Water and tissue. *Environmental Toxicology and Chemistry*, 19(8), 1951–1970.
3. Abdel-Shafy, H.I., & Mansour, M.S. (2016). A review on polycyclic aromatic hydrocarbons: source, environmental impact, effect on human health and remediation. *Egyptian Journal of Petroleum*, 25(1), 107–123.
4. Chen, S.C., & Liao, C.M. (2006). Health risk assessment on human exposed to environmental polycyclic aromatic hydrocarbons pollution sources. *Science of the Total Environment*, 366(1), 112–123.
5. Jongeneelen, F.J., Scheepers, P.T., Groenendijk, A., Aerts, L.A.V., Anzion, R.B., Bos, R.P., & Veenstra, S.J. (1988). Airborne concentrations, skin contamination, and urinary metabolite excretion of polycyclic aromatic hydrocarbons among paving workers exposed to coal tar derived road tars. *American Industrial Hygiene Association Journal*, 49(12), 600–607.

6. Ramesh, A., Walker, S.A., Hood, D.B., Guillén, M.D., Schneider, K., & Weyand, E.H. (2004). Bioavailability and risk assessment of orally ingested polycyclic aromatic hydrocarbons. *International Journal of Toxicology*, 23(5), 301–333.
7. Department of Public Health. (n.d.). Polycyclic aromatic hydrocarbons (PAHs), Cancer in Illinois Resources. Illinois Department of Public Health. http://www.idph.state.il.us/cancer/factsheets/polycyclicaromatichydrocarbons.htm.
8. Pavanello, S., Campisi, M., Mastrangelo, G., Hoxha, M., & Bollati, V. (2020). The effects of everyday-life exposure to polycyclic aromatic hydrocarbons on biological age indicators. *Environmental Health*, 19(1), 1–13.
9. Buha, A., & Williams, M. (2011). Polycyclic aromatic hydrocarbons. Retrieved October 27, 2016. https://www.healthandenvironment.org/docs/ToxipediaPAHArchive.pdf
10. Unwin, J., Cocker, J., Scobbie, E., & Chambers, H. (2006). An assessment of occupational exposure to polycyclic aromatic hydrocarbons in the UK. *Annals of Occupational Hygiene*, 50(4), 395–403.
11. NIOSHTIC-2. (2000). Detection of polycyclic aromatic hydrocarbons (PAHS) in skin wipe samples of automotive mechanic. Trainees. National Institute of Occupational Safety and Health, U.S. Department of Health and Human Services. https://www.cdc.gov/niosh/nioshtic-2/20041454.html.
12. Boeniger, M., Neumeister, C., & Booth-Jones, A. (2008). Detection of polycyclic aromatic hydrocarbons (PAHS) in skin wipe samples of automotive mechanic. Trainees. American Industrial Hygiene Conference and Exposition.
13. Agency for Toxic Substances and Disease Registry. (2009). *Polycyclic aromatic hydrocarbons (PAHs), how should patients exposed to PAHs be treated and managed?* U.S. Department of Health and Human Services Public Health Service. https://www.atsdr.cdc.gov/csem/polycyclic-aromatic-hydrocarbons/treatment_and_management.html
14. Patel, A.B., Shaikh, S., Jain, K.R., Desai, C., & Madamwar, D. (2020). Polycyclic aromatic hydrocarbons: sources, toxicity and remediation approaches. *Frontiers in Microbiology*, 11, 2675.
15. Agency for Toxic Substances and Disease Registry. (2007). *Toxicological profile for benzene.* U.S. Department of Health and Human Services Public Health Service. https://www.atsdr.cdc.gov/toxprofiles/tp3.pdf.
16. Salviano Dos Santos, V.P., Medeiros Salgado, A., Guedes Torres, A., & Signori Pereira, K. (2015). Benzene as a chemical hazard in processed foods. *International Journal of Food Science*, 1–7. https://doi.org/10.1155/2015/545640.

Chapter 2

Aromatic amines

2.1 INTRODUCTION

Production of aromatic amines (AAs) began around the mid-19th century. William Henry Perkin, in 1856 in the attempt to produce quinine from aniline sulfate, instead, produced a dark material that could dye silk a bright mauve. He was later asked by a dye company to make large quantities at such time he established operations that began the start of the aniline dye industry. Production of benzidine began around 1853. It was used in the production of several azo dyes, such as Palatine Orange and Congo Red. Unfortunately, during this time frame, human health hazards that can result from exposure to certain AAs were not known and not widely studied.

AAs constitute one of the major groups of known carcinogens. As early as 1895, a report by German surgeon Rehn surfaced indicating that four workers appeared at a clinic with bladder cancer. These workers were employed at a manufacturing plant producing magenta dye from crude commercial aniline. Three worked in the same factory, in what was known as the fuchsin room. The fourth patient with bladder cancer worked at another factory where fuchsin was also manufactured. After examining the manufacturing process, Rehn concluded that the cancers seen in these patients were a result of inhaling aniline vapor. Decades later, it was demonstrated that 2-napththylamine, benzidine, 1-naphthylamine, 4-ethoxyacentanilide, and 4-aminobiphenyl were carcinogenic in the bladder of humans.[3]

Approximately 18 years after beta-naphthylamine production started in Germany in 1898, a correlation between beta-naphthylamine exposure and bladder tumors was reported. By the early 1950s, extensive evidence existed indicating that dyestuff workers experienced a higher incidence of bladder cancer, and indications pointed to AAs used in the production of the dyes. Furthermore, other occupations that used AAs in their processes also showed increased cases of bladder cancer including workers who manufactured rubber using beta-naphthylamine, medical technicians and nurses who used benzidine to test for occult blood, and tire remolders who were exposed to vapors from AAs during tire heating procedures. In addition, noted were the patients who had received beta-naphthylamine mustard

DOI: 10.1201/9781003220114-2   15 therapy as a treatment died of bladder cancer 2 to 11 years after beginning such treatment.

Humans continue to be exposed to AAs from a variety of sources to individuals as cigarette smoke, pharmaceutical products, and diet. It has been shown that tobacco smoke contains 4-aminobiphenyl as well as other human bladder carcinogens and that smokers have an increased risk of developing bladder cancer. Many pharmaceutical products that are commonly used contain AAs, such as dapsone, procaine, p-amino salicylic acid, sulfanilamide, and minoxidil. Some AAs are present as natural components of the food consumed by humans, and others are formed during the process of cooking.[4]

AAs are substances typically identified as chemical compounds having in their molecular structure one or more aromatic rings[7] (Figure 2.1). AAs are a group of semi-volatile compounds that are highly toxic and have adverse effects on the respiratory and cardiovascular systems. In addition to causing cancers, these compounds can also exacerbate asthma conditions and increase the risk of pneumonia.[10,11]

The initiation of bladder cancer by some AAs was first connected to the workplace. Workplace exposure can still occur today during the manufacture of dyes, rubber, pharmaceuticals, and a variety of other products. N-acetylation appears to be the primary reduction mechanism for human exposure to AAs.[10]

AAs have been and are currently widely used across the globe as chemical precursors for materials to include dyes, antioxidants, and resins. Newly discovered AAs are synthesized every year. Many AAs have been listed as carcinogenic to humans.[6] AAs are used in many different industrial and agricultural activities such as antioxidants in the production of rubber and in cutting oils, as intermediates in azo dye manufacturing, and as pesticides. The usage of AAs in these industries can pose exposure potential to workers that can significantly impact health and well-being. AAs are a common

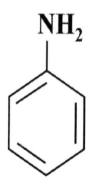

Figure 2.1 Basic aromatic amine – aniline.

contaminant in many working environments, such as in the chemical and mechanic industries and aluminum transformation.[6] There has always been interest on how the chemical structure of amines determines the biological effect ever since it has been established that certain amines are carcinogenic to humans. Gaining a better understanding of this relationship can form the basis and assist in developing a strategy that should be taken to assess hazards and the risk associated with exposure to these substances. Having this knowledge coupled with an effective risk assessment and reduction strategy will prevent exposures and disease development in humans. The common factor with the wide array of AAs is an amino group connected to an aromatic network.[12] Figure 2.2 lists some AAs and structures.

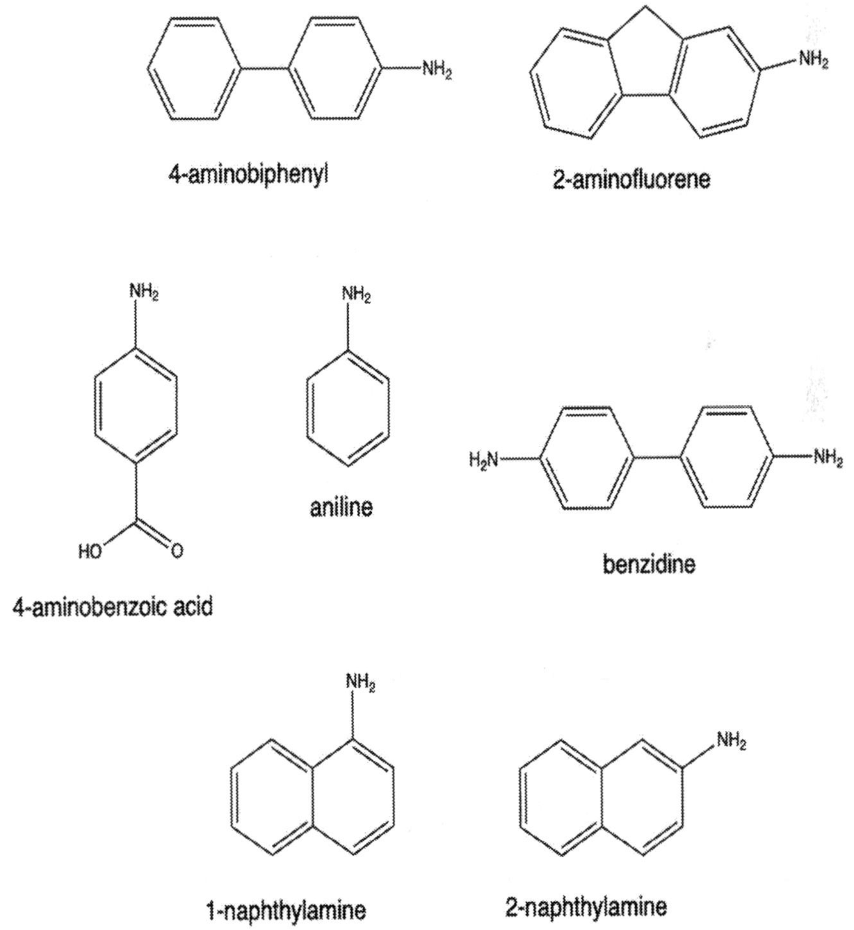

4-aminobiphenyl

2-aminofluorene

4-aminobenzoic acid

aniline

benzidine

1-naphthylamine

2-naphthylamine

Figure 2.2 Examples of aromatic amine chemical structures.[16]

Azo dyes are widely used in products and represent a large portion of the dyes used in the textile industry. They represent between 60% and 70% of all dyestuff associated with textile production. AAs can be released through dermal, systemic, and biotransformation of azo dyes. There are more than 896 azo dyes used in the production of textiles; more than 426 are potential parent compounds of one or more of the approximate 22 regulated AAs. These dyes are widely used and disposed throughout the environment.

Table 2.1 represents a list of AAs that have been identified as possessing mutagenic capability.[3] Not all mutagenic amines are listed in Table 2.1.

2.2 HETEROCYCLIC AROMATIC AMINES

Another form of AA that can create long-term health effects are heterocyclic aromatic amines (HAAs). This class of amine compounds is carcinogenic and can be formed in meats, fish, and poultry that are prepared in the average household. Some HAAs have been found in beer and wine and can be formed during condensation of tobacco smoke. More than 20 HAAs have been discovered in meats, fish, and poultry prepared under normal household cooking. These levels of HAAs can range from less than 1 ppb to more than 500 ppb, depending on the type of meat. As such, it is recognized that human exposure to HHAs can be of concern if exposed at high levels. HAAs are effectual genotoxins in bacterial and mammalian cell assays and are proven to produce strong carcinogenic effects in experimental animals. This knowledge has raised questions on the safety of foods containing HAAs and the potential impact on the health of humans when exposed. In addition, some epidemiological data exist that suggest that HAAs may contribute to some forms of cancers, such as colorectal, prostate, and breast cancers, found in humans connected to the consumption of cooked meat products. The cancer risk posed to humans is uncertain, and the risk factors that have been estimated have wide ranges.[10]

The United States Department of Agriculture (USDA) recognizes that some studies suggest that the risk of cancer may exist from eating food cooked using high heat cooking techniques. However, the USDA also states that eating moderate amounts of food consumed using this method of cooking does not pose a problem to humans currently.[15]

2.3 EXPOSURE PATHWAYS

AAs can be an exposure hazard in the workplace through inhalation, dermal contact, and ingestion. These compounds can also come in contact with human through other means such as by some food consumption and cigarette smoking. It is not clear what level of AA consumed through food

Table 2.1 Mutagenic aromatic amines

Chemical name	CASRN
Aniline	62-53-3; 142-04-1
2,2'-Dimethylbenzidine	84-67-3; 198487-76-2
2,6-Xylidine; 2,6-dimethylaniline	87-62-7
o-Aminobenzenesulfonic acid	88-21-1; 13846-13-4;
2-Nitro-4-chloroaniline	89-63-4
N,N-diethyl-p-phenylenediamine	93-05-0; 2198-58-5
2,4-Dimethylaniline	95-68-1
2-Amino-4-methylphenol	95-84-1
2-Amino-4-chloropheno	95-85-2
4,6-Dinitro-2-aminophenol	96-91-3
2,4-Dinitroaniline	97-02-9
2-Amino-5-nitroanisole	97-52-9
m-Nitroaniline	99-09-2
2,6-Dichloro-4-nitroaniline	99-30-9
2-Amino-4-nitrophenol	99-57-0
m-Nitro-o-anisidine	99-59-2
N,N-dimethyl-p-phenylenediamine	99-98-9
p-Nitroaniline	100-01-6
m-Phenylenediamine	108-45-2
2-Amino-5-nitrothiazole	121-66-4
2-Chloro-4-nitroaniline	121-87-9
2-Amino-5-nitrophenol	121-88-0
p-Aminoacetanilide	122-80-5
p-Aminophenol	123-30-8
p-Phenetidine	156-43-4
1,2,4-Triaminobenzene (dihydrochloride	615-47-4; 615-71-4
2-Methyl-p-phenylenediamine; 2,5-diaminotoluene	615-50-9; 95-70-5; 25376-45-8
p-Phenylenediamine (dihydrochloride)	624-18-0; 106-50-3
6-Methoxy-2-benzothiazolamine	1747-60-0
2-Bromo-4,6-dinitroaniline	1817-73-8
1,4-Naphthalenediamine	2243-61-0
1,1-Bis(4-aminophenyl)cyclohexane	3282-99-3
6-Chloro-2,4-dinitroaniline	3531-19-9
4-Nitro-m-phenylenediamine	5131-58-8
2,5-Diaminoanisole	08-02-5307
2-Amino-6-nitrobenzothiazole	6285-57-0
4-Amino-2,5-dimethylaniline	07-01-6393
3-Amino-5-nitro-2,1-benzisothiazole	14346-19-1
4,6-Diaminoresorcinol (dihydrochloride)	16523-31-2; 15791-87-4
p-Anisidine (hydrochloride)	20265-97-8; 104-94-9

can contribute to the development of cancers in humans. However, the flag is raised, and studies have indicated that there is a possibility of cancer forming mostly supported by tests performed using animals. In the workplace, several AAs can also be absorbed percutaneously. Table 2.2 provides a list of some AAs, their origin, and potential impacts to humans and the environment. Having knowledge of the AAs that can potentially produce cancerous effect in humans and the industry of use provides needed information on how to develop an exposure risk assessment control program and process to protect workers while in the work environment. Assistance in determining the potential exposure pathway, developing an exposure assessment and control program for the workplace is found in Attachment A.

AAs are used as intermediates in the synthesis of azo dyes, and AA exposures from consumer-developed products is a real risk to human health that can yield mutagenic or carcinogenic effect from some AAs. Therefore, many AAs have been banned from usage in various countries.[3]

2.3.1 Inhalation exposures and impact

A widely used component of various industrial processes, AAs can be hazardous to workers during various production activities. As such, steps must be taken to protect workers from exposures to these products. As a part of providing protection to workers, the work environment must be evaluated, the risk of exposures must be known, and methods to eliminate or reduce exposures employed. Inhalation exposures should be evaluated prior to beginning any work activities. In addition, the development and performance of a task analysis, developing a monitoring strategy and conducting a personal air sampling campaign should be conducted. The result gained from evaluating the tasks performed by workers will determine the following:

- If the task can be performed safely.
- If additional protective measures are necessary.
- If protective equipment is needed.

   AAs are sampled in air typically by two methods:
- Method 1 consists of the gradual filter of air through 0.1 M HCl solution to trap AAs as hydrochloride salts. Then the AAs are removed using chloroform containing bis-2-ethylhexylphosphate (BEHPA) and derivatized, before being subjected to gas chromatography-mass spectrometry analysis, with isobutyl chloroformate.
- Method 2 used samples collected in adsorbent tubes and extraction by thermal desorption with any other treatment and derivatization.[10]

Table 2.2 Some aromatic amines' potential origin and impact[3]

Chemical name	Origin	Potential environmental impact	Chemical structure
Aniline	Isocyanates, rubber, dyes, explosives, pesticides, and pharmaceutical manufacturing processes	Volatile organic compounds (VOCs) have the potential to form ozone, toxic to aquatic life; potentially carcinogenic and genotoxic	
4-Chloroaniline	Oil refining; tobacco smoke; forest fires; manufacture of dyes, pesticides, and other chemicals	Toxic to humans	
Toluene 2,4-diamine	Manufacture of toluene diisocyanate (for elastomers), dyes, resins, fungicides	Carcinogenic and genotoxic, VOC	
2-Naphthylamine	Manufacture of dyes	Toxic to humans and aquatic life; Potential carcinogenic and genotoxic	
4,4'-Methylenebis (2-chloroaniline)	Polyurethane manufacturing	Carcinogenic, VOC	
4,4'-Methyl-ene dianiline	Manufacture of polyurethanes, dyes, and epoxy resin	Recalcitrant adsorption onto particulate matter; toxic to humans and aquatic life; possibly carcinogenic and genotoxic; VOC	
N-Nitrosodi-phenylamine	Dye manufacturing, pharmaceuticals, and rubber production	Harmful to humans	
Benzidine	Manufacturing of dyes	Toxic to humans – carcinogenic	
2-Aminobiphenyl	Cigarette smoking	Carcinogenic and genotoxic	
2-Amino-1-methyl-6-phenylimidazo [4,5-b] pyridine	Cooking of meats	Carcinogenic and genotoxic	

2.3.2 Dermal exposures and impact

There are few studies detailing the relationship between lesions found on the skin and the percutaneous uptake of substances in workers who are exposed to hazardous substances. A study conducted by Korinth, G., Weiss, T., Angerer, J., and Drexler, H. in 2006, confirmed that dermal absorption of AAs is still of concern for exposed workers and that internal exposure to AAs can increase in workers with impaired epidermal barriers. The skin condition and the type of skin protection used and care can affect the uptake of AA through the dermal route.[6]

This 2006 study focused on and place emphasis on the relevance of dermal absorption of the AA *ortho*-toluidine to four workers with different skin conditions. The workers worked in the automobile industry, and the evaluation focused on exposures that occurred during the hardening process of rubber products. Skin attributes considered included the following:

- Whether or not the skin was healthy?
- Was the skin erythematous and burned skin?
- Were there any appearance of pompholyx eczema?

The results of the case report also showed that dermal absorption of *o*-toluidine through damaged epidermal barrier can be significantly higher than through skin that was believed to be healthy.[4]

Contact sensitivity to AAs has been recognized for decades. Allergic responses to aromatic amino compounds have been found from clinical experiences and animal assays.[5] Dermal exposure is a concern for many chemicals; the same is true for AAs. These chemicals can cause skin conditions ranging from mild to serious in exposed individuals. Absorption of AAs through percutaneous can result from direct dermal contact, airborne vapor, aerosols, and particulates. Little information and knowledge about the absorption rate of chemical via impaired skin is known. The consideration for impaired skin is important because for example in the printing industry it has been reported that more than 70% of workers have impaired skin on their hands. In this case, the breakthrough time for gloves worn is important to provide adequate protection and limit the potential for skin absorption to take place.[6]

Naphthylamines such as N-phenyl-l-naphthylamine and N-phenyl-2-naphthylamine have been shown to be skin sensitizers. The contact dermatitis resulting from sensitivity to naphthylamines has not been reported frequently as such there are only few references regarding sensitization to naphthyl compounds. Some of these references include low incidence of possible sensitization to naphthyl mix (N-phenyl-2-naphthylamine and di-(2-naphthyl)-p-phenylencdiamine), which has been documented by Amsterdan van Ketel and Bieber & Foussereau. They reported findings of six patients with sensitization to N-phenyl-2-naphthylamine. In addition, acne, leucoplakia, and hypersensitivity to sunlight have been associated with prolonged periods of occupational exposure to N-phenyl-2-naphthylamine. Some AAs constitute a potential hazard with respect to contact dermatitis because of their alkaline nature. The principal dermatological response to AAs is an allergic contact dermatitis.[5]

2.3.3 Ingestion exposure and impact

Many AAs have been reported to be toxic and mutagenic and can often cause carcinogenesis in animals and, in some cases, humans. Exposure of the general population to AAs is mostly because of tobacco smoking and ingestion of the heterocyclic amines that are formed during the cooking of meat and fish products (particularly if charcoal-broiled). The mode that these AAs exert their toxic and carcinogenic effects is currently unclear and is still being studied.[9] However, according to the USDA, moderate consumption of food consumed using this method of cooking does not pose a problem to humans.[14] Therefore, ingestion of AAs during food consumption will not be considered as an ingestion hazard of concern in this section.

Ingestion of AAs is not a well-documented event. In fact, ingestion of these products is more of a concern in work environments where workers are consuming food and beverages in work areas where these chemicals are present, poor hygiene practices are present, and housekeeping in the work environment is lacking. Some questions to consider when evaluating the potential for ingestion of AAs in the workplace are as follows:

- Is handwashing stations available to sanitize prior to leaving the work area?
- Are food items consumed in the chemical handling and work processing area?
- Are chemical products and processes located near eating areas?

In the work environment, careful considerations must be given to the chemicals that are used and whether ingestion is a possible route of entry for workers. Appendix A contains tools that can be used to perform an assessment of what conditions can present a potential ingestion hazard for AAs during usage in a process.

2.4 CHEMICAL TO HUMAN INTERFACE

AAs are of concern globally because they represent one of the most important classes of industrial and environmental chemicals that are used in abundance and that can exert carcinogenic and mutagenic effects on humans. A variety of AAs have been designated as potent carcinogens and mutagens and/or hemotoxicants. The challenge is that exposure to AAs is not confined to just one industry; in fact, AAs occur in different industrial and agricultural activities as well as in tobacco smoking. To exert their carcinogenic effect, AAs must be metabolized to reactive electrophiles. This reaction typically involves an initial N-oxidation to N-hydroxy arylamines and N-hydroxyarylamides.[1] More than one in eight of known or suspected human carcinogens are attributed to AAs or chemicals that can convert into one, rendering AAs as a significant class of human carcinogens.[12]

It has been established that certain AAs are carcinogenic in humans, and questions continue to surface on how the chemical structure determines the biological effects. The common denominator is the cancer-causing impact, and the amino group bound to an aromatic system. The chemical reactivity of the amino group is significantly dependent on the mesometric interaction with the aromatic system. Based on existing knowledge, many if not all AAs have the potential to produce cancers in humans.[13] AAs can also contaminate the ambient air from environmental tobacco smoke. There is increasing evidence that some bladder cancers in smokers can be attributed directly to AAs.[8]

Exposure to AAs can occur in different industrial and agricultural processing activities. Compounds which are considered human carcinogens include, but not limited to, 2-naphtylamine and 4-aminobiphenyl; they were once used as dye, antioxidant, or in the rubber industry. AAs have been found to be present in low quantities in many occupations including aluminum production, hairdressing, painting, and shoemaking. The extent to which these compounds contribute cancer observed in these occupations has not been extensively examined. AAs require metabolic activation to transform into fully carcinogenic agents.[2]

Several AAs had been identified as having the risk of producing cancer which easily can enter the water system via discharged effluent and degradation depending on their solubility in water. Metabolic activation of the amino group determines the acute and chronic toxicity of the compounds.[1] The basicity of the amine is significantly reduced by the aromatic ring because of its electron-withdrawing properties. This effect can be either strengthened or weakened depending on the type of substituents connected to the ring and on the nitrogen group.

AAs are produced from azo dye decomposition that are extensively used and/or produced in industries such as textiles, plastics, cosmetics, food colorants, rubber processing, pesticide production, polymers, surfactants, corrosion inhibitors, and in tobacco smoke. AAs may also be discharged into the environment from power generators, many manufacturing processes, improper waste disposal, unintentional spills, and conversion of azo dyes to the parent AAs through bacterial action. AAs can pass through the bloodstream during inhalation of smoke produced from cigarettes (smokers or second-hand smoke) where they will be filtered out from the blood and discharged in the urine. As such, there is the risk for the AAs to contact the bladder cells and potentially trigger the development of cancer cells.[1]

2.5 EXPOSURE CONTROL METHODS

Controlling exposures to AAs continues to evolve. In fact, over decades, many AAs have been banned from usage once it has been discovered that they are among the cancer-causing groups. Controlling exposures in an occupational environment is of primary concern as these products are used in so many industries to produce or enhance a wide array of products and services. Methods that are of typical use in controlling exposures to chemicals are as follows:

- Conducting exposure assessment (Appendix A) to determine exposure mode and implement methods to eliminate or control exposures while protecting workers
- Collecting air sample (Appendix A) and compare the results to the occupation exposure limit as noted by the Occupational Safety and Health Administration (OSHA) and/or the American Conference of Governmental Industrial Hygienist (ACGIH)
- Product elimination and substitution
- Engineering control
- Personal protective equipment (PPE)
- Administrative controls

These methods should be explored when developing a control strategy that will eliminate or control worker exposure to AAs.

2.5.1 Product elimination and substation

Many countries have discontinued the use of some AAs. For example, over the past decades, several manufacturers in the United States have discontinued synthesizing benzidine-based dyes and pigments and have replaced them with phthalocyanine, o-toluidine (2), 3,3'-dimethoxybenzidine, phenylenediamine, and dioxyazine-type dyes. Although 2-naphthylamine can be used as an intermediate in the production of dyes and antioxidants, its use is now generally prohibited or severely restricted. In addition, 2-amino-1-naphthalene sulphonic acid is normally substituted in dye manufacturing, and the antioxidant N-phenyl-2-naphthylamine is usually prepared from the condensation of aniline and 2-naphthol rather than from 2-naphthylamine.[5]

There are many more AA products that have been removed from usage across the globe. The United States is not the only country banning the use of AAs when information is gained on health effects that can lead to health hazards including mutagenic or carcinogenic impacts on humans. There is a potential that some AAs that are banned in one country may be allowed for use in another. Therefore, careful consideration and review should be given before using these products.

2.5.2 Engineering controls

When seeking to control exposures to AAs, it is important to take advantage of deploying the most effective control measures. Engineering control when designed properly offers the best means to control exposure to AAs. Designing closed-loop systems that will allow chemical addition and handling through mechanical means offer the greatest protection and the best way to reduce or eliminate the potential for the development of cancers and contact dermatitis among employees who are associated with exposures to AAs. Let's consider as an example of where engineering control will be useful in processes where AAs are produced or used. Adding chemicals to the process is a good task where engineering control is helpful by adding exhaust ventilation systems that are effective in removing fumes away from workers and the work environment. The use of fume hoods where chemicals are used in a small amount during laboratory analytical and research procedures is another effective way to use engineering controls when handling AAs. Additional information on methods to control workplace exposures to AAs can be found in Appendix A.

2.5.3 Administrative controls

This control method is not advisable to be used to control exposures to AAs simply because there has not been an established time limit recommendation that has been deemed safe for exposure that will not increase health risk. Administrative controls include a wide array of practices and documentation including procedures, work orders, signs, rotation of work assignments, and training.

2.5.4 Protective equipment and effectiveness

PPE can be effective in protecting workers from exposure to AAs during various processing activities. For example, laboratories and employers have developed and implemented procedures that are used for various chemical handling tasks to protect workers. PPEs that can be used to protect workers when working with AAs include gloves (chemical gloves), chemical aprons, safety glasses, and respiratory protection. When deploying PPE for personal protection, it is important to ensure that the PPE is appropriate to provide the level of protection needed. Appendix C has been designed to assist in the selection of the right PPE for the hazard encountered.

2.6 HUMAN BIOLOGICAL IMPACT

Several AAs had been identified as having the risk of producing cancer. These substances are of industrial importance and, therefore, found in many workplaces. In addition, these substances can enter the water system through discharged effluent and degradation depending on their solubility in water. Several AAs that are commonly detected in cigarette smoke include o-toluidine, 2-napthylamine, benzidine, 4-aminobiphenyl, and 4,40-methylenebis(2-chloroaniline). These amines are considered as influential factors for the occurrence and development of lung and urinary bladder cancer in humans. It has been reported that the risk of suffering from bladder cancer from transitional cell carcinoma (TCC) because of tobacco smoke can increase with the duration of smoking and number of cigarettes smoked.[1]

Many studies over decades have been conducted to determine the parental AAs and their metabolites in biological specimens including plasma, urine, hemoglobin, hair, bile, and milk produced by humans to evaluate the potential exposure of AAs in humans. However, it has been found that the most convenient method is the collection and analysis of the urine because of the opportunity to collect a large number of samples.[1] Some prominent AAs that are known carcinogens are listed in Table 2.3.

2.6.1 Potential impact to offsprings

There has been some positive reporting that AAs can have significant impacts on the fetus. However, a preponderance of the evidence found addressed damage to non-human embryos. Nevertheless, there have been indications that AAs are also harmful to the liver of a human fetus. According to T Aune, A Haugen, and E Dybing, 1985, human fetal liver microsomes have been found to metabolize some AAs.[16] Having the limited knowledge of human embryo effects, we must proceed with extreme caution to prevent harmful effects to offsprings. The best way to avoid fetus exposure to AAs is to protect people from becoming exposed while at work. The workplace at this point represents the most prominent place where people could be exposed to significant levels of AAs. Employer's play, a pertinent role in the plight to prevent exposure, therefore, prevent the possibility of exposure and damage to the fetus.

Table 2.3 Carcinogenic aromatic amines

Compound	Chemical abstract services
4-Aminobipheynl	92-67-1
Benzidine	92-87-5
4-Chloro-o-toluidine	95-69-2
2-Naphthylamine	95-59-8
o-Aminoazotoluene	97-56-3
5-Nitro-o-toluidine	99-55-8
4-Chloroaniline	106-47-8
2,4-Diaminoanisoe	615-05-4
4,4'-Diaminodiphenylmethane	101-77-9
3,3'-Dichlorobenzidine	91-94-1
3,3'Dimethoxybenzidine	119-90-4
3,3'Dimethylbenzidine	119-93-7
3.3'Dimethyl-4,4'-diaminodiphenylmethane	838-88=0
p-Cresidine	120-71-8
4,4'-Methylene-bis-(2-chloro-analine	101-14-4
4,4'-Oxydianiline	101-80-4
4,4'-Thiodianiline	139-65-1
o-Toluidine	95-53-4
2,4-Toluylendiamine	95-80-7
2,4,5-Trimethylaniline	137-17-7
2-Methoxyaniline	90-04-0
4-Aminoazobenzene	60-09-3
2,4-Xylidine	95-68-1
2,6-Xylidine	87-62-7
Aniline	62-53-3

2.7 TREATMENT OF RESULTANT ILLNESS

The most prominent illness realized from exposure to AAs in the industrial or work environment in humans is bladder cancer. However, AAs found in tobacco smoke are also carcinogenic to humans and can cause cancers in the lungs, bladder, kidneys, pancreas, esophagus, larynx, and pharynx.[1] Some potential methods of treating illnesses and diseases caused by AAs are shown in Table 2.4. These treatment methods are based on the treatment advocated by the Mayo Clinic and can be located on their website.[15] Treatment that may be recommended by the medical community depends on the type and stage of the disease during the time of discovery, and the patient preference.

Table 2.4 Treatment for aromatic amine disease and illnesses

Illness	Diagnostic method	Symptoms	Potential treatment methods
Cancer – bladder	Cystoscopy – examining the inside of the bladder using a scope, biopsy, urine cytology (examining a urine sample), imaging test	Symptoms may include blood in urine, frequent and/or painful, urination, and back pain	Surgery, chemotherapy, immunotherapy, and radiation therapy
Cancer – esophagus	Barium swallow study, using a scope to examine the esophagus (endoscopy), collecting a sample of tissue for testing (biopsy)	Signs include difficulty swallowing, weight loss, chest pain, pressure of burning in the chest, indigestion or heartburn that gets worse, and coughing or hoarseness, coughing, or hoarseness	Surgery, chemotherapy, radiation therapy, combined chemotherapy and radiation therapy, targeted drug therapy, and immunotherapy
Cancer – kidney	Blood and urine tests, imaging tests, and biopsy (removal of tissue sample)	Symptoms include blood in urine, pain in the back or side that doesn't go away, loss of appetite, weight loss, tiredness, and fever	Surgery, non-surgical treatment (heat or freeze cancer cells), targeted therapy, immunotherapy, and radiation therapy
Cancer – throat	Using a scope (endoscope or laryngoscope) to get a closer look at your throat, remove a sample to test tissue (biopsy), and imaging test (CT, MRI, PET)	Symptoms may include, a cough, sore throat, changes in your voice (hoarseness or not able to speak clearly), difficulty swallowing, pain in the ear, a lump or sore that doesn't heal and loss of weight	Radiation therapy, surgery, chemotherapy, target drug therapy, immunotherapy, rehabilitation on completion of treatment, and palliative (supportive) care
Cancer – pancreas	Imaging tests that create pictures of internal organs, blood test, using a scope to create ultrasound pictures of the pancreas, and biopsy	Symptoms may include abdominal pain that radiates to the back, loss of appetite, weight loss, yellowing of the skin and the whites of your eyes (jaundice), light-colored stools, dark-colored urine, itchy skin, recent diagnosis of diabetes or existing diabetes that's becoming more difficult to control blood clots, and fatigue	Surgery, chemotherapy, radiation therapy, clinical trials, and palliative care
Cancer – lungs	Imaging test, sputum cytology, tissue sample (biopsy)	Signs and symptoms of lung cancer may include a cough that doesn't go away, coughing up blood, shortness of breath, chest pain, hoarseness, headache, loss of weight, and bone pain	Surgery, radiation therapy, chemotherapy, palliative care, immunotherapy, targeted drug therapy, stereotactic body radiotherapy, chemotherapy, and radiation therapy
Contact dermatitis	Patient interview and patch test	Symptoms include a red rash, severe itching, dry, cracked scaly skin, bumps, and blisters, sometimes with oozing and crusting, and swelling, burning, or tenderness	Steroid creams or ointment and oral medications

REFERENCES

1. Benigni, R. (2005). Structure-activity relationship studies of chemical mutagens and carcinogens: Mechanistic investigation and prediction approaches. *Chemical Reviews, 105*, 1767–1800.
2. Boon, Y.H., Zain, N.N.M., Mohamad, S., Fauzid, H.M., Alias, Y., Chandrasekaram, K., Rahim, N.Y., Yahayaa, N., & Raoov, M. (2019). Determination of aromatic amines in urine using extraction and chromatographic analysis: A minireview. *Analytical Letters, 52*(18), 2974–2992. https://doi.org/10.1080/00032719.2019.1636057
3. Bruschweiler, B.J., & Merlot, C. (2017). Azo dyes in clothing textiles can be cleaved into a series of mutagenic aromatic amines which are not yet regulated. *Regulatory Toxicology and Pharmacology, 88*, 214–226.
4. Committees on Amines Board of Toxicology and Environmental Health Hazards Assembly of Life Sciences National Research Council, Aromatic amines: *An assessment of the biological and environmental effects*, National Academy Press, Washington D.C., 1981.
5. Korinth, G., Weiss, T., Angerer, J., & Drexler, H. (2006). Dermal absorption of aromatic amines in workers with different skin lesions: A report on 4 cases. *Journal of Occupational Medicine and Toxicology, 1*, 17. https://doi.org/10.1186/1745-6673-1-17.
6. Korinth, G., Weiss, T., Penkert, S., Schaller, K.H., & Angerer, J. (2007). Percutaneous absorption of aromatic amines in rubber industry workers: Impact of impaired skin and skin barrier creams. *Occupational and Environmental Medicine*, London, 64(6), 366–372. https://doi.org/10.1136/oem.2006.027755.
7. Inoue, N. (2017). Analysis of six aromatic amines stability in workplace measurement, work environment research group. National Institute of Occupational Safety and Health, Tama-Ku, Kawasaki.
8. Vineis, P., & Pirastu, R. (1997). Aromatic amines and cancer. *Cancer Causes & Control, 8*(3), The Harvard-Teikyo Program Special Issue, 346–355.
9. Shuker, L. K, International Council of Scientific Unions. Monitoring and Assessment Research Centre & International Programme on Chemical Safety. (1986). *The Health effects of aromatic amines : a review*, L. K. Shuker et al. (Ed.), Monitoring and Assessment Research Centre. https://apps.who.int/iris/handle/10665/40073.
10. Turesky R, J. (2002). Heterocyclic aromatic amine metabolism, DNA adduct formation, mutagenesis, and carcinogenesis. *Drug Metabolism Reviews, 34*(3), 625–650.
11. Lucaire, V., Schwartz, J.-J., Delhomme, O., Ocampo-Torres, R., & Millet, M. (2018). A sensitive method using SPME pre-concentration for the quantification of aromatic amines in indoor air. *Analytical & Bioanalytical Chemistry, 410*(7), 1955–1963.
12. Shuang, W., Daniel, H., Sugamori K.S., & Denis, M. (2019) Primary aromatic amines and cancer: Novel mechanistic insights using 4-aminobiphenyl as a model carcinogen. *Pharmacology & Therapeutics* 200, 179–189.

13. World Health Organization International Agency For Research on Cancer, *IARC Monographs on the Evaluation of Carcinogenic Risks to Humans, Volume 99; Some Aromatic Amines, Organic Dyes, and Related Exposures,* Lyon France, 2010.

14 Ye, X., Peng, Y., Niu, Z. et al. Novel approach for the rapid screening of banned aromatic amines in dyed textiles using a chromogenic method. (2018). *Analytical and Bioanalytical Chemistry 410,* 2701–2710. https://doi.org/10.1007/s00216-018-0941-x.

15. What are heterocyclic aromatic amines? (usda.gov)

16. Aune, T., Haugen, A., & Dybing, E. (1985). Metabolism and mutagenicity of aromatic amines by human fetal liver. https://doi.org/ 10.1007/BF00343124.

Chapter 3

Nitrosamines

3.1 INTRODUCTION

N-Nitrosamines are a class of chemical compounds having the essential feature of N-nitroso compounds, which are the N–N=O structure. The R1 and R2 groups attached to the amine nitrogen can range from a simple hydrogen atom to more complex chemical substituents such as ring structures that contain the nitrogen atom. Exposures to the human population to nitrosamines can result from the formation of N-nitroso compounds in food during storage or preparation or in the body, generally in the stomach (Figure 3.1).

Individual nitrosamines are not found in seclusion; however, they occur in mixtures of various nitrosamines. Nitrosamines and their antecedents occur in a wide variety of foods and manufactured and natural products including agricultural chemicals, tobacco, detergents, rust inhibitors, cutting fluids, rubber additives, solvents, drugs, plastics, tanned leather products, textiles, and cosmetics. Generally, nitrosamines are not added purposely to foods or consumer products; they are formed from constituents of the foods or products that are either naturally present, such as the amines that are part of the structure of proteins in meat, or added during production, for example, the nitrates or nitrites added to meats as preservatives. Nitrosamines are formed when nitrites react with a secondary or tertiary amine. The concentration of nitrosamines typically increases over time, and their formation is enhanced by high temperatures, such as while frying food, and high acidic environments or solutions such as in the acid in the stomach. The formation of nitrosamines can be inhibited by ascorbic acid and its isomers. Ascorbic acids and these isomers are often added to food to prevent the formation of nitrosamine.

N-Nitrosamines exposure can also occur from the nitrosamines produced internally in the digestive tract. Approximately 5% of ingested nitrates are reduced to nitrites in saliva, and these nitrites can react in solution with secondary and tertiary amines, N-substituted amides, carbamates, and other related compounds to form N-nitroso compounds within the gastrointestinal tract. The formation of nitrosamine is considered a major source of exposure.[5]

DOI: 10.1201/9781003220114-3

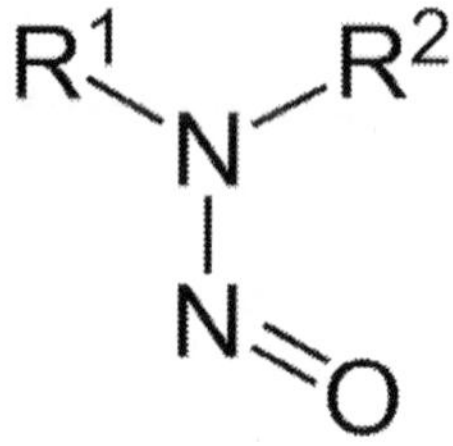

$$R^1 \diagdown \underset{\underset{O}{\overset{\displaystyle N}{\parallel}}}{N} \diagup R^2$$

Figure 3.1 Nitrosamine structure.

Nitrosamines occur as contaminants in different food categories such as vegetables, cheese, drinking water, milk, and fish. The primary carcinogenic nitrosamines that are found in food throughout the world are nitrosodimethylamine (NDMA), N-Nitrosodiethylamine (NDEA), N-Nitrosopyrolidine (NPYR), N-Nitrosopiperidine (NPIP), and N-Nitrosodibutylamine (NDBA). Nitrosamines can also be found in food products, wine, beverages, and tobacco products. These NAs are formed in some food products from naturally occurring amines that are present in food. The amines are nitrosated by either nitrite or nitrogen oxides. They can induce tumors in a wide variety of organs. In addition to the externally produced sources of exposure to N-nitroso compounds, there is evidence that such substances are produced internally at various sites in the human body after ingestion or inhalation of precursors.[1]

Nitrosamines can be formed by the reaction of secondary amines with nitrite-derived nitro-sating species, such as NO_2, N_2O_3, and N_2O_4, under specific acidic conditions. Some of these N-nitrosamines are found only in tobacco formed by the reaction of reactive nitrosating species derived from nitrite with their corresponding tobacco alkaloids.[4] The four major tobacco-specific nitrosamines (TSNAs) are N-nitrosonornicotine (NNN), N-nitrosoanatabine (NAT), N-nitrosoanabasine (NAB), and 4-methyl-N-nitrosamino1-(3-pyridyl)-1-butanone (NNK). TSNA may be present at very low levels in growing tobacco or freshly harvested green tobacco. The bulk of nitrosamines are formed during the post-harvest treatments, for example, during curing and ageing.[4] Table 3.1 contains some nitrosamines along with usage and production status. This table is not comprehensive and does not represent all nitrosamine-based products.

Formation of NDMA and other volatile N-nitrosamines in drinking water may involve at least the following factors:

- They are primarily formed during disinfection using chloramines.
- Formation of NDMA is a slow process.
- Reactions with ion-exchange resins having quaternary amine functional groups can yield NDMA.

Table 3.1 Nitrosamines

Nitrosamine	CAS no.	Usage	Commercial production
N-Methyl-N'-nitro-N-nitrosoguanidine	70-25-7	Used currently as research with no known commercial usage	Not produced commercially in the United States
N-Nitrosodi-n-butylamine	924-16-3	Primarily in research. Previously used as an intermediate in the synthesis of di-n-butylhydrazine	Not produced commercially in the United States. Detected in a variety of products resulting from nitrosation of amines
N-Nitrosodiethanolamine	1116-54-7	Primarily in research. No known commercial use	Not currently produced commercially in the United States
N-Nitrosodiethylamine	55-18-5	Primarily in research. Some previous use includes as an additive in gasoline and lubricants, antioxidant, stabilizer in plastics, and fiber industry solvent	Not currently produced commercially in the United States
N-Nitrosodimethylamine	62-75-9	Used primarily in research. Previously used as an intermediate in the electrolytic production of 1,1 dimethylhydrazine	Not currently produced commercially in the United States
N-Nitrosodi-n-propylamine	621-64-7	Used in small quantities in laboratory research. No known commercial use	Never produced in commercial quantities
N-Nitroso-N-ethylurea	759-73-9	Used to synthesize diazoethane in laboratory and studied for promoting plant growth	Not produced for commercial use in the United States
4-(N-Nitrosomethylamino)-1-(3-pyridyl)-1-butanone (NNK)	64091-91-4	Laboratory experimental usage	Not currently produced commercially

(Continued)

Table 3.1 (Continued) Nitrosamines

Nitrosamine	CAS no.	Usage	Commercial production
N-Nitroso-N-methylurea	684-93-5	Once used to synthesize diazomethane in laboratory and replaced by other reagents. Small quantities used in the study	Never been produced commercially in the United States
N-Nitrosomethylvinylamine	4549-40-0	Used as a research chemical	No evidence that ever produced in the United States commercially. Available in small quantities for research only
N-Nitrosomorpholine	59-89-2	Used as a research chemical	No evidence of commercially produced in the United States
N-Nitrosonornicotine	16543-55-8	Research chemical	No evidence of production in the United States
N-Nitrosopiperidine	100-75-4	Research chemical	Prepared in 1863 by the reaction of nitrogen dioxide on piperidine. Numerous patents have been issued; however, no evidence of commercial manufacturing
N-Nitrosopyrrolidine	930-55-2	Research chemical	Not produced commercially in the United States
N-Nitrososarcosine	13256-22-9	Limited research use	No evidence of production in the United States

- Reactions with ozone (O_3) and chlorine dioxide (ClO_2) can produce NDMA under specific circumstances, but these agents may also reduce NDMA formation when used as pre-oxidants that can destroy NDMA precursors.
- Nitrosamines undergo decomposition by ultraviolet light and other processes.[3]

3.2 EXPOSURE PATHWAYS

Humans are exposed to N-nitrosamines primarily through smoke emitted from cigarettes and from occupational sources.[6,16] Some exposure to nitrosamines can also occur through the consumption of some food products. These exposure sources can lend itself to exposure through all potential means of contact for humans.

3.2.1 Inhalation exposure and impact

Nitrosamines can become an exposure issue for workers in the work environment primarily through inhalation. Inhalation of dust and fumes represents the majority exposure pathway for these products that have been determined to present health hazards. The workplaces that have been evaluated appear to be the environment with significant nitrosamines exposure in the production of rubber and tires. Exposure to nitrosamine in the rubber and tire industry has been measured, and the result shows high levels of N-nitrosomorpholine (NMOR) and N-nitrosodimethylamine. These high levels are found in the air in the production areas, associated shops, and storage rooms.[7,14]

3.2.2 Dermal exposure and impact

Nitrosamines and its products have the potential to cause dermal adverse effects once it comes in contact with the skin. The National Institute for Occupational Safety and Health (NIOSH) was informed by one of its grantees, that they had confirmed the presence of nitrosamine and diethanolnitrosamine in commercial cutting fluids produced by various companies. Cutting fluids are used in machine operations such as drilling, gear cutting, grinding, lathing, and milling.[10,17] While using these products, care must be given to avoid skin contact. To avoid or limit the potential for dermal contact, the worker should wear the appropriate gloves and keep all skin surfaces covered to provide a barrier between the worker and the source. The worker should keep work area surfaces clean and orderly and wash hands frequently especially after leaving the work area and before eating.

3.2.3 Ingestion exposure and impact

Ingestion of food seems to be a significant contributor to potential exposure to nitrosamines when food comes in contact with the human digestive system through ingestion. Although exposure may occur through ingestion, the exposure in most cases may not be considered harmful as food product contents are monitored by the FDA.[14] However, further reduction in consumption can be achieved when the individual monitors his food intake and reduces his consumption of food products containing nitrosamines.

3.3 CHEMICAL TO HUMAN INTERFACE

The primary organs affected by N-nitrosamines are the liver, kidney, bladder, esophagus, stomach, intestine, nose, skin, tongue, brain, and nervous system. Tissues can also be impacted; however, they are dependent on the structure of the compound, the dosage, route of administration, and exposed animal species.[1] A preponderance of the literature written on nitrosamines indicates that carcinogenic and health risks and effect as it pertains to humans rest heavily on animal research. For example, volatile N-nitrosamines are a class of chemical contaminants that exhibit unambiguous, genotoxic carcinogenicity in rodent bioassays. Therefore, the International Agency for Research on Cancer classified them as probable human carcinogens as does the US Environmental Protection Agency (EPA).[3] The lack of data with regards to human exposure and health effect is, perhaps, directly related to the limited use of these chemicals within the manufacturing community and the small amount used in research.

3.4 EXPOSURE CONTROL METHODS

Trace amounts of nitrosamines in food may be potentially harmful to health. To reduce the total human exposure to nitrosamines and the lowest effective dose, it is advisable to reduce exposure to nitrosamines in the daily intake to the smallest dose feasible.[1] For example, it is known that nitrates are added to a variety of cured meats such bacon to prevent spoilage and toxin formation from *Clostridium botulinum* as well as to improve the flavor and appearance of the meat. Fried bacon is one of the most significant dietary sources of nitrosamines because the reaction of nitrite with amines and amino acids in the meat occurs during cooking resulting in volatile nitrosamines.[2] Foods which have been shown to contain volatile nitrosamines include cured meats, primarily cooked bacon, beer, some cheeses, nonfat dry milk, and at times fish.[8]

Having this knowledge, a reduction of these food items in the diet would be a great way of reducing potential exposures from that source. The FDA issues guidance and recommendation of step manufacturers of drug products should take to discover and prevent unacceptable levels of nitrosamine in pharmaceutical products.[11]

Exposure to nitrosamines in the work environment is limited based on usage and the fact that many of these chemicals are not being manufactured in the United States. As such, potential exposure to workers is minimized through limited availability and use. When these products are used or produced in the work environment, a full-scale exposure assessment and risk analysis (Appendix A) should be conducted before they are used or produced.

3.4.1 Product elimination and substitution

Nitrosamines have been detected in the food products consumed, the drugs used to resolve health issues as well as used in the work environment. Product elimination can be useful to eliminate exposures in areas where nitrosamines are produced or used. However, there may be times when substitution is not feasible such as in some processes where nitrosamine is a byproduct of the process. However, a full attempt should be made by exploring if elimination and substitution are feasible. In the production of rubber and rubber products, the occurrence of nitrosamines during the process depends on their formation from used vulcanization accelerators based on amines and the presence of nitrosating agents, such as diphenylnitrosamine, and nitrous gases, in products or production areas. Elimination of one or both precursors for nitrosamine formation resulted in a significant reduction of air contamination of nitrosamines.[14] When elimination or substitution is not feasible, in such case, other methods must be considered and used for the protection of workers and the environment. Products that are used by consumers are eliminated from use continuously as more information is discovered indicating the presence of nitrosamines at levels that can be harmful to human health. For example, in 2021, the FDA alerted health care professionals that the drug Chantix was being recalled due to the presence of unacceptable levels of nitrosamine.[13]

3.5 ADMINISTRATIVE CONTROLS

Training and education represent commonly used administrative controls that can be useful in controlling exposures to nitrosamine when it comes to consumer usage as well as in the work environment. Educating consumers on the potential health effects of these products and the type of food or medication that may contain these products provides consumers with a choice on how to approach the use of various products. Administrative controls that can be useful in the work environment include procedures, work policies, task rotation among employees, utilizing a robust work planning and control process, and modification of tasks to reduce hazards and worker contact with the task.

3.6 PERSONAL PROTECTIVE EQUIPMENT

Personal protective equipment (PPE) can be effective only if the appropriate PPE is selected and maintained. PPE is useful to protect workers from being exposed to nitrosomine products. In addition, users are trained on proper usage, limitations, and maintenance of the equipment. Respiratory protection can be an effective PPE when trying to protect workers from airborne exposures. Careful selection of these devices is necessary because airborne exposures can manifest themselves by way of fumes as well as particulates. In addition, gloves and clothing designed to cover the exposed skin could prevent skin exposures. More information on the types and selection of PPE is found in Appendix C.

3.7 HUMAN BIOLOGICAL IMPACT

Exposure to high concentrations of nitrosamines is associated with increased mortality from cancers of the esophagus, oral cavity, and pharynx. The International Agency for Research on Cancer (IARC) classified several nitrosamines as carcinogens or potentially carcinogenic to humans. The IARC considers NDMA and NDEA as probable carcinogenic to humans, and NDBA, NPIP, and NPYR as potentially carcinogenic to humans.[1,12]

The study has demonstrated an association between exposure to N-nitrosamines and shorter average telomere length in peripheral blood. Telomeres are specialized DNA-protein structures that are located at the end of eukaryotic chromosomes. Telomeres are essential regulators of cellular life span and play an important role in maintaining the stability of chromosomes. Short telomere length in peripheral blood appears to be a risk marker for cardiovascular diseases and breast, bladder, head and neck, lung, and renal cell cancers in humans.[15]

Table 3.2 contains some nitrosamines along with health effects and potential exposure mechanism. The potential health effects are based primarily on animal research. This table is not comprehensive and does not represent all nitrosamine-based products.

3.7.1 Potential impact on exposed individual

N-Nitrosamines are among the most powerful and adaptable carcinogens that are capable of being formed from a wide range of precursors that are widely disbursed throughout the environment such as nitrogen-containing substances and nitrosating agents that are in various living organisms. N-Nitrosamines are a large group of genotoxic chemical carcinogens that are found in human diet and other environmental media. These compounds are toxic, mutagenic, and teratogenic and can also be transported across the placenta. The various nitrosating agents that are found in water, air, and food products along with amines and other nitrogen-containing substances are precursors for the formation of nitrosamines.[1]

N-Methyl-N'-nitro-N-nitrosoguanidine (MNNG) is an N-nitrosamine alkylating agent that is a yellow crystal at room temperature. The substance is soluble in water, dimethyl sulfoxide, and polar organic solvents. MNNG reacts violently with water and with some nucleophiles such as amines and

Table 3.2 Nitrosamines exposure impact

Nitrosamines	Health effect	Exposure
N-Methyl-N-nitro-N-nitrosoguanidine (MNNG)	Based on animal studies: cancerous tumors in animals have been detected in studies (stomach, liver, lung blood vessels)	Unknown. However, it is presumed limited to scientific usage in a laboratory setting. Potential exposures to researchers
N-Nitrosodi-n-butylamine	Based on animal studies: carcinogen effects – bladder upper digestive tract, liver, stomach	Potential exposures to researchers
N-Nitrosodiethanolamine	Based on animal studies: cancer of the liver, kidney, and nasal cavity tumors	Known contaminants of cosmetics, lotions, shampoos, cutting fluids, certain pesticides, antifreeze, and tobacco. Formed within these products by reactions of precursors that include nitrosating agents and amines
N-Nitrosodiethylamine	Based on animal studies: cancerous tumors in animals, liver, kidney respiratory tract, upper digestive tract, lung, and trachea	Unknown exposure may occur from food, tobacco, beverages, tobacco smoke, drinking water, and air pollution. Nitrosamine produced frequently during rubber processing and may be present as contaminants in the product made with rubber. Exposure potentials are dependent on nitrosamine migration from the product to the body
N-Nitrosodimethylamine	Based on animal studies: cancerous tumors of the liver, respiratory tract, kidney, blood vessels, bile duct, lung, and nasal cavity	May have some exposure in laboratory research work. Unknown exposure quantities in food, water smoking beverage, tobacco, herbicides, drinking water, air pollution, and drugs formulated with aminopyrine
N-Nitrosodi-n-propylamine	Based on animal studies: malignant tumors of the lung, nasal and paranasal cavities, esophagus, and liver	Found in rubber products, cheese, alcoholic beverages, herbicide trifluralin. May be formed in the upper gastrointestinal tract following ingestion of food containing nitrates and secondary amines. May occur in cigarette smoke

(Continued)

Table 3.2 (Continued) Nitrosamines exposure impact

Nitrosamines	Health effect	Exposure
N-Nitroso-N-ethylurea	Based on animal studies: malignant tumors found in areas of the body such as the nervous system, kidneys, lymphoreticular system, liver, bone, skin, blood vessels, and stomach	Human exposure may occur through food, water, air, and formation in the human body from precursor ingested, and smoking tobacco
4-(N-Nitrosomethylamino)-1-(3-pyridyl)-1-butanone	Based on animal studies: carcinogenic tumors liver, lungs nasal cavity, and blood vessels	Potential exposure from smoking tobacco and people exposed to side-stream smoke. May be formed in the mouth during tobacco chewing or use of snuff
N-Nitroso-N-methylurea	Based on animal studies: carcinogenic effect includes the nervous system, kidney, colon, and blood vessels	Potential exposure from air and water, laboratory analysis
N-Nitrosomethylvinylamine	Based on animal studies: cancer of the tongue, pharynx, nasal cavity, and esophagus	Potential exposure from research activities
N-Nitrosomorpholine	Based on animal studies: cancerous tumors found in areas such as liver, bile duct, respiratory tract, and nasal cavity	Potential exposure among laboratory workers or workers in the tire manufacturing industry
N-Nitrosonornicotine	Based on animal studies: cancers in the nasal cavity	Potential exposures from tobacco products. Produced during the curing of tobacco from secondary amines
N-Nitrosopiperidine	Based on animal studies: cancers of the nasal cavity, esophagus, lung, and stomach	Potential exposures from food, tobacco product consumption, and the resultant smoke
N-Nitrosopyrrolidine	Based on animal studies: cancer of the liver	Produced when food is preserved with contaminated nitrite. Potential exposures during cooking, and vapor is inhaled and food consumption
N-Nitrososarcosine	Based on animal studies: cancer of the esophagus, liver, and nasal cavity	Formed when nitrite preserved foods containing secondary amines are heated. Exposure can occur during inhalation of vapors during cooking or ingesting prepared food

can explode when exposed to heat or on high impact. At low pH, MNNG releases nitrous acid, and at high pH in the presence of hydroxyl alkalis, it produces diazomethane a highly toxic gas and when heated to decomposition MNNG emits nitrogen oxides.

MNNG is suspected to be a human carcinogen based on evidence of carcinogenicity obtained from studies in experimental animals. Studies in animals to MNNG showed the presence of cancerous tumors in several animal species during experimentation in several different tissues and by several different routes of exposure. MNNG caused tumors primarily at the site of administration, mostly in the gastrointestinal tract, including tumors of the stomach, small intestine, and large intestine.

The data available from human cancer studies are insufficient to evaluate the relationship between human cancer and exposure to MNNG. The data obtained from evaluating three deaths resulting from brain tumors and one death resulting from colon cancer were reported by workers in a genetics laboratory for more than 13 years. The four subjects had been exposed to MNNG from 6 to 15 years prior to death and had worked around other types of carcinogenic materials.[5] Evidence directly linking exposure to nitrosamines and cancer in humans has been somewhat difficult to establish. The reason is that exposures in the human environment often occurred to low concentrations of nitrosamines and complex mixtures. As such, only two specific N-nitroso compounds, NDMA and NDEA, have been classified by the International Agency for Research on Cancer (IARC) as "probably carcinogenic to humans." [9]

3.7.2 Potential impact to off-spring

The literature on nitric oxide (NO) has not documented the general description of the relative activity of the NO synthase system during fetal and early postnatal development in humans. However, there are many studies of NO synthase and responsiveness to a variety of specific stimuli in fetal, young, and adult animals.[3]

3.8 TREATMENT OF RESULTANT ILLNESS

More than 90% of the 300 tested nitrosamines have shown potent carcinogenic effects in experimental studies in at least 39 different species and many different organs. The sites where tumors were found in animals include oral cavity, stomach, bladder, and brain. The highest concentrations of nitrosamines that humans can be exposed to in the environment occur and have been measured in the rubber industry.[9] Treatment of symptoms and diseases resulting from exposure to nitrosamines will be based on relevant therapeutic strategies as evaluated by the attending physicians. To determine a specific treatment will depend on the medical history of the worker. The type of treatment will be determined at the onset of symptoms and clinical evaluations and tests done and will more than likely follow the treatment protocols currently followed for such symptoms and diagnosed disease.

REFERENCES

1. Bhangare, R.C., Sahu, S.K., & Pandit, G.G. (2015). Nitrosamines in seafood and study on the effects of storage in refrigerator. *Journal of Food Science Technology* 52, 507–513. https://doi.org.zproxy.proxy.library.oregonstate.edu/10.1007/s13197-013-1016-9
2. Cancer Risk from Nitrosamines in Pork Bacon Prepared by: Risk Assessment and Analytics Staff Office of Public Health Science Food Safety and Inspection Service U.S. Department of Agriculture February 20, 2014.
3. Hrudey, S.E., Bull, R.J., Cotruvo, J.A., Paoli, G., & Wilson, M. (2013). Drinking water as a proportion of total human exposure to volatile N-Nitrosamines, *Risk Analysis*, 33(12), 2179–2208. https://doi.org/10.1111/risa.12070
4. Li, Q., Krauss, M.R., & Hempfling, W.P., 2006, Wounding of Root or Basal Stalk Prior to Harvest Affects Pre-harvest Antioxidant Accumulation and Tobacco-specific Nitrosamine Formation during Air Curing of Burley Tobacco (Nicotiana tabacum >), Journal of Agronomy and Crop Science, 192 (4), 267-277. https://doi.org/10.1111/j.1439-037x.2006.00217.x.
5. Report on Carcinogens Profile: Nitrosamines; 15th RoC 2021 (nih.gov).
6. Sedlo. I., Kolonić. T., Tomić. S. (2021). Presence of nitrosamine impurities in medicinal products. *Arh Hig Rada Toksikol*, 72(1), 1–5. doi:10.2478/aiht-2021-72-3491.
7. Scanlan, RA. (1983). Formation and occurrence of nitrosamines in food. *Cancer Res.*, 43(5 Suppl), 2435s–2440s.
8. Straif, K., Weiland, SK., Bungers, M., et al. (2000). Exposure to high concentrations of nitrosamines and cancer mortality among a cohort of rubber workers. *Occupational and Environmental Medicine*, 57,180–187.
9. Updates on possible mitigation strategies to reduce the risk of nitrosamine drug substance-related impurities in drug products, FDA.
10. https://www.fda.gov/regulatory-information/search-fda-guidance-documents/control-nitrosamine-impurities-human-drugs.
11. Control of Nitrosamine Impurities in Human Drugs Guidance for Industry U.S. Department of Health and Human Services Food and Drug Administration Center for Drug Evaluation and Research (CDER) February 2021 Pharmaceutical Quality/ Manufacturing Standards/ Current Good Manufacturing Practice (CGMP) Revision 1, 41–166.qxd (who.int)
12. FDA Updates and Press Announcements on Nitrosamine in Varenicline (Chantix) I FDA,

13. Spiegelhalder, B., & Preussmann, R. (1983). Occupational nitrosamine exposure. 1. Rubber and tyre industry. *Carcinogenesis, 4*(9), 1147–1152. doi: 10.1093/carcin/4.9.1147.

14. Li, H., Jonsson, B.A., Lindh, C.H., Albin, M., & Broberg, K. (2011). N-nitrosamines are associated with shorter telomere length. *Scandinavian Journal of Work, Environment & Health, 37*(4), 316–324. https://doi. org/10.5271/sjweh.3150.

15. Fritschi, L., Benke, G., Risch, H. A., Schulte, A., Webb, P.M., Whiteman, D.C., Fawcett, J., & Neale, R. E. (2015). Occupational exposure to N-nitrosamines and pesticides and risk of pancreatic cancer. *Occupational and Environmental Medicine, 72*(9), 678–683. http://www.jstor.org/stable/43870306.

16. National Institute of Occupational Safety and Health (1976). Current Intelligence Bulletin 15: Nitrosamines In. National Institute of Occupational Safety and Health, U.S. Department of Health and Human Services. https:// www.cdc.gov/niosh/docs/78-127/78127_15.html.

17. Fajen, J.M., Fine, D.H., Rounbehler, D.P. (1980). N-Nitrosamines in the factory environment. *IARC Scientific Publications*, (31), 517–30.

Chapter 4

Azo dyes

4.1 INTRODUCTION

Dyes have increasingly become a necessity for product manufacturers across many industries to add aesthetics to the presentation of the goods produced as such there can be hundreds to thousands of variations available for usage from time to time. They are used in many products manufactured that are consumed in high quantities by consumers around the world. Dyes are instrumental in adding or enhancing the color of various products to make them more appealing to eyesight. Not only are dyes important to the manufacturing industry, they also have important usage in areas such as the food and scientific community. When used in scientific applications, it can be helpful in illuminating or as a detection mechanism for various constituents. When used in food, it provides or enhances the color and the appearance of the food item rendering the product more pleasing to the eyesight and providing the appearance of freshness.

More than 800,000 tons of dyes are produced each year across the globe, and 60%–70% of these dyes are classified as azo dyes. Synthetic dyes are used in textile, dyeing, paper, printing, color photography, pharmaceutical, food, cosmetics, and other industries. Table 4.1 offers a more expansive list of potential usage of these dyes. Azo dyes represent about one-half of all the dyes in common use across the globe and are employed as coloring agents in the textile, food, and pharmaceutical industries.[5,6]

A dye is generally defined as a colored organic compound or mixture that is used for adding color to an object such as paper, plastic, cloth, leather, or solution. During early years, dyes were produced from natural constituents that presented some limitations in preparation and usage. The extractions were found to be unsuitable for large-scale preparation because of the labor and time involved in preparation, and a small amount of dye could be obtained through processing a large amount of vegetable or animal matter. This process of producing dyes was not only expensive, but also it yielded a product that lack uniformity and reproducible quality. Specifically, they were subjected to damages or fading due to light. As such, natural dyes have been replaced largely with synthetic dyes. The azo dyes constitute the

DOI: 10.1201/9781003220114-4

Table 4.1 Azo Dye Product Usage Examples

Textiles	Sportswear, swimsuits, undergarments, hats, bedding, towels
Leather products	Purses, wallets, briefcases, chair covers, pouches, leather jackets, footwear
Accessories	Wristwatch bands, gloves, hairpieces, hats, sleeping bags, wigs
Materials that encounter food products	Plastic lunch boxes, colored napkins, bakery bags
Cosmetics	Hair dyes, nail polish, napkins
Scientific	Analytic processes

Table 4.2 Classification of Azo Dyes

Dye type	Characteristics
Acidic azo dyes	Characterized based on the presence of an acidic or phenolic group. The acid group makes the dye more soluble.
Basic azo dyes	Characterized by the -HK2 or -NR2 group.
Direct or substantive azo dyes	Characterized due to the ability to directly dye the fibers without the need for a mordant.
Ingrain azo dyes	Water-soluble dyes that are formed on the fiber by dipping the fabric in an alkaline solution mixture.
Mordant azo dyes	When fixed on the fiber, these dyes have the capability to form complex metallic compounds after treated with metallic salts.
Synthetic fiber dyes	Used to dye synthetic fibers such as terylene, nylon, and rayon.
Stilbene-azo dyes	Consists of yellow and orange direct dyes used to color cellulosic fibers.

largest group of all synthetic colorants produced. These dyes have a colorimetric system having an azo group, -N+N-, in conjunction with one or more aromatic systems where there can be one or more azo group present in the dye molecule. There are many azo dyes, and their structures vary and can be further classified into many categories; seven of these categories are shown in Table 4.2.[12,13] These classifications can be further expanded into more than 70 categories as shown in Table 4.3.

Many of the more than 70 categories of azo dyes are well categorized and studied; therefore, much is known about their impact on humans and the environment. Some of the dye categories are still being explored to further categorize potential health effects. These categories are listed in Table 4.3. Tables 4.2 and 4.3 serve only as examples and not a comprehensive list because of the constant changes in dye production. Dyes having more than one or more azo group account for the largest family of organic dyes. These types include acid dyes for polyamide and protein substrates to include nylon, silk, and wool; disperse dyes (DDs) for substances classified as being hydrophobic such as polyester and acetate and reactive and direct dyes for cellulosic substrates that include paper, rayon, cotton, and linen.[4]

Table 4.3 Azo Dyes Categories

Common name	Key information
Acid orange 5	Used as a pH indicator
Acid orange 7	Dying wool
Acid orange 19	Color polyamide and wool fibers
Acid red 13	Produced as a sodium salt
Acid red 88	Used to dye cotton red
Alcian yellow	Used primarily in along with toluidine blue to assist in the visualization of *Helicobacter pylori*
Alizarine yellow	Primarily used as a pH indicator
Allura red AC	Used as a food dye
Amaranth (dye)	Applied to natural and synthetic fibers, paper, leather, and some resins
Amido black 10B	Used in biochemical research for staining of protein on transferred membrane blots and in criminal investigations to detect the presence of blood
Aniline yellow	Used in microscopy for staining, in pyrotechnics for yellow-colored smoke, ink-jet printers, insecticides, varnishes, waxes, oil stains, styrene resins, intermediate synthesis for other dyes
Arylide yellow	Used primarily as industrial colorants for plastics, paints and inks, acrylics, and watercolors
Azo violet	Used commercially as a violet dye and experimentally as an indicator of pH
Azorubine	Primarily used in foods that are treated with heat after fermentation
Basic red 18	Used for coloring textiles
Biebrich scarlet	Used to color hydrophobic materials like fats and oils
Bismarck brown Y	Used to stain acid mucins, mast cell granules, and cartilage in bone specimens
Black 7984	Used as the tetrasodium salt. Use in food has not been permitted in the United States, European Union, Australia, and Japan.
Brilliant black BN	Used in food decorations and coatings. E151 has been banned in the United States, Japan, and Switzerland. Approved for use in the European Union.
Brown FK	*Used as a food dye. Not approved for used in the European Union, Australia, Austria, Canada, United States, Japan, Switzerland, New Zealand, Norway, and Russia*
Brown HT	Use as a food dye.
Calconcarboxylic acid	Used as an indicator for complexometric titrations of calcium
Chrysoine resorcinol	*Formerly used as a food additive. Banned in Europe in 1977 and in the United States in 1988. Used as a pH indicator*
Citrus red 2	Used to color the peel of an orange
Congo red	*Not used primarily because of its carcinogenic properties*

(Continued)

Table 4.3 (Continued) Azo Dyes Categories

Common name	Key information
D&C red 33	Used as a colorant in mouthwashes, cosmetics, and hair dyes
Direct blue 1	Used as a substantive dye for textiles with high contents of cellulose
Direct blue 15	Useful for dying cotton and other cellulose materials
Disperse orange 1	Useful in conducting experiments with flash photolysis
Eriochrome black T	Used in complexometric titrations
Evans blue (dye)	Useful in physiology in estimating the proportion of body of water contained in blood plasma, as a viability assay and to assess the permeability of the blood brain barrier to macromolecules
Fast yellow AB	*Once used as a food dye. No longer used as toxicological data has shown it is harmful.*
Orange 1	One of the seven original food dye
Hydroxynaphthol blue	Used to determine the endpoint in complexometric titrations
Janus green B	Used to stain mitochondria supravitally
Lithol rubine BK	Used as a food dye and as a standard magenta in color printing processes
Metanil yellow	Used as a pH indicator
Methyl orange	Used as a pH indicator
Methyl red	Used as a pH indicator
Methyl yellow	Used as a dye for plastics and as a pH indicator
Mordant brown 33	Used as a dye for wool fabric
Mordant red 19	Used in textile dyeing
Naphthol AS	Used as coupling mechanism in the preparation of some azo dyes
Oil red O	Used for staining of neutral triglycerides and lipids
Orange B	Food dye. Approved in the United States for use only in hot dog and sausages surfaces
Orange G	Used in histology for many staining formulations
Orange GGN	*Formerly used as a food dye. No longer used in food*
Para red	Use to dye fabric
Pigment yellow 10	Used as a yellow colorant, as yellow road marking on highways in the United States
Ponceau 2R	Used for staining in histology
Ponceau 3R	*Once used as a red food dye*
Ponceau 4R	*Can be used in a variety of food products, used in Europe, Asia, and Australia. Not been approved for human consumption by the United States*
Ponceau 6R	It is used as a food dye and in histology to stain fibrin
Ponceau S	Used to prepare stains for rapid detection of protein on nitrocellulose membranes

(Continued)

Table 4.3 (Continued) Azo Dyes Categories

Common name	Key information
Prontosil	Used as a drug
Red 2G	Used as a dye for coatings, inks, paper, food dye (in Europe). Ban in countries such as Australia, Canada, Israel, Japan, Norway, and Malaysia
Scarlet GN	No longer permitted for use in food or ingested drugs and may only be used in externally applied drugs and cosmetics.
Sirius red	Primarily used in staining methods for collagen and amyloid; in diagnostic such as to observe fibrosis inflammation from cancer
Solvent black	Used in the staining of neutral triglycerides and lipids
Solvent red 26	Primarily use is as a fuel dye
Solvent yellow 56	Used as a fuel dye in the European Union and used to dye products such as hydrocarbon solvents, waxes, fat, oil, shoe polishes, colored smoke, and mineral oil
Solvent yellow 124	Used as a fuel dye in the European Union
Sudan I	Used to color various food items, oils, waxes, solvents, polishes, and colored smoke
Sudan II	Used to color nonpolar substances such as oils, fats, waxes, greases, and hydrocarbon products. Banned from usage as a food dye due to toxicity
Sudan III	It is used to color nonpolar substances, for example, fats, oil, greases, waxes, some hydrocarbons, and fuel dye
Sudan IV	Used to color substances such as oils, fats, greases, waxes, some hydrocarbons, and staining lipids, triglycerides, and lipoproteins
Sudan red 7B	Used in biology for staining, in industries as a fuel dye, and can be present in laser toners
Sudan red G	Used for coloring of fats, oils, and waxes, temporary tattoos, hair dyes, polystyrene, cellulose, synthetic lacquers, and as a food dye.
Sudan stain	Used as dye for various plastics and used to stain biological samples
Sudan yellow 3G	Used as a pigment for cosmetics and printer toners, dye in inks used in printers, in pyrotechnics used to produce yellow smoke
Sunset yellow FCF	Used in foods, condoms, cosmetics, and drugs
Tartrazine	Primarily used for food coloring, cosmetics, and medication
Tropaeolin	Used as a pH indicator
Trypan blue	Dye for cotton textiles and is used as a stain to color dead tissues or cells
Yellow 2G	Food coloring

A variety of azo dyes are used primarily in the process of adding color to food products. Coloring food is not only to make the food products visually aesthetic but also to reinstate the original appearance that may have been lost during the production process. Many countries have discontinued the use of many of the azo dyes in food because of known and suspected health effects on humans. Suppliers use the color of food products as one of the important factors by which they evaluate the quality of food products. As such, consumers use the same factor in choosing whether a product is acceptable for purchase and consumption. A variety of food colorants that are used are of both natural and synthetic origin. The challenge is that most of the dyes obtained from natural sources are unstable and often undergo degradation easily during food processing. Thus, synthetic dyes are attractive for use and are widely used because of their stability and lower production cost.[2] The dye products that are no longer used or are classified as known carcinogens are highlighted in Table 4.3. Note that other dyes may have been banned or classified as known carcinogens as knowledge becomes available concerning impacts and health effects.

There are more than 75 known categories of azo dyes, and a percentage of these dyes are not currently used because of known toxicological effects on the human body. The list represented in Table 4.2 is not inclusive, and some dyes may have been omitted. Nevertheless, azo dyes still represent or account for more than 50% of dye usage when it pertains to providing aesthetics to products, diagnostic procedures, etc.

Although widely used across the globe, azo dyes are not as widely as seen in previous years used as an additive for coloring food products to make them visually aesthetic and to reinstate the original appearance lost during the production process because of known health effects. Some countries have prohibited the use of many azo dye products in food applications.[3]

The Food and Drug Administration (FDA) provides guidance on dye usage as it pertains to the food supply. A list of approved dye substances that are approved is published by the FDA for current usage in food products.[14] This list can change as frequently as knowledge is gained into the viability of the various dyes and knowledge gained on health effects.

Many of the food dyes that are listed in Table 4.4 are banned for usage in food in many countries inconsistently. Table 4.5 shows examples where these dyes may be banned in one country and yet can be used in others. Thus, it is pertinent that companies that produce dyes domestically and internationally know and understand the laws of the various countries concerning azo dyes they produce and use. As information and knowledge are gained by the various countries on the impact of these dyes on the human population, decisions are being made on whether a particular dye continues to be safe for human consumption; therefore, the list in Table 4.5 continues to change.

Attention must be paid to the regulations promulgated and enforced by each country when considering usage of the many azo dyes and developing

Table 4.4 Azo Dyes used in Food

Name	Color
Allura red (E129)	Red
Amaranth (E123)	Reddish-brown, dark red to purple
Brilliant black BN (E151)	Black
Brown FK (E154)	Brown
Brown HT (E155)	Brown
Citrus red 2 (E121)	Orange to yellow or dark red
Carmoisine (E122)	Red to maroon
LitolRubine BK (E180)	Red
Orange B	Orange
Sunset yellow (E110)	Yellow
Tartrazine (E102)	Lemon yellow

Table 4.5 Examples of Food Dye Inclusion for Some Countries

Country	Approved for use	Banned from use
Australia	E102, E110E122, E124, E129, E151, E155	E121, E123, E154, E180, orange B
China	E102, E110, E122, E123, E124, E129,	E121, E151, E154, E155, E180, orange B
EU	E102, E110, E122, E123, RE124, E129, E151, E154, E180	E121, E155, orange B
India	E102, E110, E122, E124,	E121, E123, E129, E151, E154, E155, E180, orange B
Japan	E102, E110, E123, E124, E129,	E121, E122, E151, E154, E155, E180, orange B
New Zealand	E102, E110E122, E124, E129, E151, E155	E121, E123, E154, E180, orange B
United States	E102, E110, E121, orange B	E122, E123, E124, E129, E155, E154, E155, E180,

a risk-based program to ensure workers are protected from exposures and adverse health effects. Appendix A can be helpful in developing risk assessments for these dyes.

4.1.1 Azo dye production

Azo dye manufacturing mostly consists of five basic steps that include diazotization, coupling, isolation-filtration, drying, grinding, and standardization.[9] These basic steps are shown in Table 4.6.

These five steps serve as a basis for what it takes to make one of the most common dye classes. Each dye family has its own set of processes and tests it undergoes to create the end color.

Table 4.6 Basic Azo Dye Manufacturing Process

Step	Process	Process description
One	Diazotization	This step begins by producing a diazonium salt to react with a coupling component. The process involves nitrosation of primary aromatic amines. This is accomplished by adding a strong acid such as HCl and sodium nitrite, with ice to control the temperature. After the reaction, the product can be tested with starch iodide paper and create a blue reaction.
Two	Coupling	To complete the synthesis of an azo dye, the diazonium salt reacts as an electrophile with an electron-rich coupling component, like a phenol or an aniline. This is done through an electrophilic aromatic substitution mechanism. This reaction is controlled by monitoring pH, volume, time, and temperature throughout the process.
Three	Isolation and filtration	In the isolation-filtration process, adjust the coupling liquid for pH, salt content, volume, and temperature. Then, filter the coupling. This process separates the liquid from the dye salt crystals in the coupling liquid. The isolation-filtration stage produces a paste the consistency of thick mud. It contains at least 50% water, that will be removed by drying.
Four	Drying	A common way to dry the paste is to spread it onto trays and place it in a tray dryer. After spreading, the trays go on a rack dryer and dry at a specific temperature. Once dry, the product is grinded it to a specific particle size.
Five	Dye testing	The last step is to test the dye to ensure it meets specifications of the product. These include concentration, shade, solubility, and often use-specific tests. Testing is a moment of truth demonstrating whether or not the products meet the specified standard.

4.2 CHEMICAL TO HUMAN INTERFACE

The use of azo dyes can cause direct effects from exposures and at times indirect effects on humans and the health of the environment resulting from the discharge of industrial effluents that contain these toxic components. Studies have shown the genotoxic effects of various azo dyes. The acute toxicity of azo compounds is typically low; however, some dyes have been linked with cancer in humans. Furthermore, several studies have indicated that some of these compounds can cause DNA damage.[6] Many documented cases of azo dyes causing adverse effects in animals and humans have been studied and documented.

It has been established by a diversity of researchers and medical professionals that some azo dyes are hazardous and can cause cancer when it is metabolized in the body. For example, some dyes are metabolized to

Table 4.7 Azo Dyes Metabolized to Benzidine

Cohort description	Organ impacted	Number of participant
Workers of a tannery production plant	Bladder, leukemia, lymphomas	9,365 (7,085 men; 2,280 women
Male workers employed in tanneries	Bladder	2,926 (males)
Workers weighing and formulating or dyeing at textile printing and dyeing plants	Bladder	1,210 (1,060 men, 150 women)
Workers at a tannery	Bladder, lymphoma	1,244 (870 men, 374 women
Workers with the habit of licking brushes dipped in dye	Bladder	200 (men)
Benzidine-based dye manufacturing	Bladder	403

Source: From ref. [4].

benzidine, which is a known carcinogen. Table 4.7 provides some examples of where azo dyes have caused cancer through metabolization to benzidine.

Several azo dyes contain benzidine, which is a known human carcinogen, and these dyes can be metabolically released.[1] Azo dyes can be reduced to azoreductase enzyme, which releases aromatic amines in the bacteria of the intestines and cells of the liver.[15] Another point to consider is that bacterial azoreductase enzymes can reduce azo dyes into amines. The resultant amine metabolites have been found to be, in many cases, carcinogenic, posing a threat to human health and contributing to environmental pollutants resulting from industrial processing.[10]

Azo dyes can be harmful to humans when they come in contact with humans through all exposure routes and have been documented as such. Therefore, it is critical that when working in the dye-producing industry, the employer is cognizant of exposure potential and ensures that dye design and production systems are installed with the highest level of engineering control, employees are knowledgeable of potential hazards, they have the appropriate personal protective equipment (PPE), and training to understand and aid in risk mitigation, personnel monitoring program to ensure healthy worker, and management engagement and assume responsibility for plant operation. These dyes can also be harmful to consumers if they come in contact with a product that is produced with a chemical that is deemed to be harmful, and the chemical is capable of leaching from the product. For this reason, many azo dyes have been banned from usage in food products.

4.3 EXPOSURE PATHWAYS

There has been some link between synthetic dyes and adverse health effects seen in humans. The exploration of potential health impacts of dyes on humans continues to be a topic researched and explored. The California legislative being concerned of the possible connection between synthetic food dyes and the exacerbation of symptoms related to attention-deficit hyperactivity disorder (ADHD) in children, requested an assessment an assessment by the Office of Environment Health Hazard Assessment (OEHHA). The assessment conducted through an extensive review of the literature on synthetic dyes concluded that current literature provided a substantial amount of evidence that consuming synthetic food dyes is associated with adverse health effects in children. In addition, their findings concluded that sensitivity to synthetic dyes varies from individual to individual and that some children are more likely to be more adversely impacted by these dyes than others.[7]

Azo dyes based on its use present more of a problem when ingested and from some products encountered through skin exposure. Based on the prevailing literature, there are three main routes of exposure to azo compounds. These routes include skin absorption exposure from clothing and potential ingestion experienced by young children through sucking of the dyed material. Occupational exposure can occur in the manufacturing setting primarily through the inhalation route, dermal, and less likely ingestion.

4.3.1 Inhalation exposures and impact

The dye manufacturing production process presents many opportunities for potential exposure to workers. Inhalation hazards can be of concern while performing various tasks throughout the process. Not only is inhalation of dye product of concern but also many chemicals used at the various stages in the production process. The diversity of chemicals is dependent on the type of dye being produced. When evaluating the risk of exposure to workers during the dye production and testing process, careful considerations must be given to every chemical used in the process, their reaction to each other, and the areas in the process where a worker could possibly become exposed. As such, the risk assessment process used to evaluate inhalation exposures must be extensive and should be done by someone who understands the process inputs and outputs (Appendix A). Inhalation as a route of entry has not been identified as an issue for dye consumers.

4.3.2 Dermal exposures and impact

Contact allergy to dyes has been documented widely throughout the literature on dye production and health effects. The most common dye sensitizer studied belongs to the category of dispersed dyes. These dyes are planar and fat-soluble molecules having an anthraquinone or azo structure. Contact with dyes used in the textile industry can result in contact dermatitis.[11]

DDs accounting for more than 60% of all DDs cause most of the textile-related allergic contact dermatitis. These dyes are traditionally used for dyeing synthetic fabrics made from fibers that comprise polyester, acetate, and nylon, or a blend of these with other fiber types. In contrast, these dyes are not used to dye natural fibers such as wool, cotton, and linen, and they do not bond chemically to the fibers, and their small, lipophilic molecules can migrate easily onto the skin of the person who is wearing the garment through rubbing and exposure to water.[8]

Allergic reactions to dyes have widely been documented revealing that the most prevalent dye sensitizers belong to the disperse group of dyes. Contact allergy to these dyes can result in characteristic dermatitis. A study was conducted to evaluate contact allergy to textile dyes using a population of 3,325 cohorts. The test population consisted of 58.4% females and 41.6% males. The concluding results demonstrated that there is a high propensity for dermatitis to develop on exposure to these dyes.[11]

4.3.3 Ingestion exposure and impact

Ingestion of azo dye products does not seem to be a prominent route of entry although ingestion of these products is possible under the right environmental condition during production. Although not recognized as a prominent route of entry in a production environment, it is still a viable route of entry if workers are eating in the work environment or fail to practice good hygiene. Therefore, to prevent and avoid ingestion of chemicals and dye products, the employer must act with strict rules on where food consumption can take place and ensure that these areas are free of chemicals and cleaned frequently.

4.4 EXPOSURE CONTROL METHODS

Controlling exposures to potential and known hazards is always challenging especially when dealing with production of products that are not always stable or production processes that present an opportunity for workers to come in direct contact with the physical operation that may include chemical and physical stressors. In instances of internal exposure from dyes that have been added to food products the only control method that is effective is substitution or use avoidance. To assist with use avoidance, regulatory agencies have banned the use of many dyes in food products.

Azo dyes are viewed as being resistant to aerobic conditions and can be readily reduced by intestinal flora to form aromatic amines that have the capability to cause frequent headaches. Because of the associated problem and other more severe health issues, many countries have banned the use of most azo dyes in products that would be consumed as food.[2] Consumers can also limit the use or consumption of products that may impact their health.

4.4.1 Product elimination and substitution

When designing dyes, careful consideration of the chemical component and the impact on the human body is required. During design of these dyes, the concept of product elimination and substitution takes on a slightly different yet similar concept. When considering the dye product itself, it is important that creation does not add yet another hazardous product that can harm human health or the environment. Some dyes have been banned from usage because of their impact on human health and the environment. Product elimination is key as much knowledge is gained into the hazard potential of these dyes to avoid negative impact to consumers' health, and substitution of less hazardous chemicals in the dye-producing process is key in reduction or elimination of hazards to workers during the production stage.

4.4.2 Engineering controls

Azo dyes are among the chemical compounds that are organic in nature, and a variety of chemicals can be used at the various stages of production. The goal of employers and health and safety advocates is to avoid contact with the chemical components used and the product itself by personnel. As such, engineering controls could be beneficial to prevent exposure to workers during the manufacturing process. Effective engineering controls can include, for example, enclosures, laboratory hoods, local exhaust ventilation, or other mechanical devices that are designed to reduce chemical substances or contamination in the air.

4.4.3 Administrative controls

The wide use of administrative controls beyond training, procedures, and policies is not a viable option in controlling exposure to azo dyes because there is not enough known about the specific length of exposure that will produce cancers in humans. However, some have speculated, and others have reported years to decades before, that cancer is discovered in humans who had exposures for various lengths of time. In some cases, the latency period for cancer development is known, and in many cases, the latency period is unknown, and the length of exposure is known as shown in Table 4.8. Because of the uncertainty of when cancer can and will develop in humans, it is not reasonable in many cases to rely on administrative control methods.

4.4.4 Protective equipment and effectiveness

Protecting workers against the hazard associated with azo dye production is an important element in an effective exposure control program. Although protective equipment is the least desired line of defense because of the potential stress that it can add to the individual, it can be effective in the arsenal

Table 4.8 Azo Dyes Exposure Time Reported

Cohort description	Organ impacted	Exposure
Workers of a tannery production plant	Bladder, leukemia, lymphomas	Latency – 15+ years
Male workers employed in tanneries	Bladder	Latency <15 years
Workers weighing and formulating or dyeing at textile printing and dyeing plants	Bladder	Employed >1 year
Manufacture of magenta dye	Bladder cancer	Employed for at least 6 months
Manufacture of fuchsin (dye)	Bladder	Employed 15–29 years

Source: From ref. [4].

of ensuring workers are protected against some workplace hazards. When designing dye production processes, the goal should be to engineer out as many hazards that are feasible. When all fail when it comes to protecting workers, that is when PPE is necessary. In such a situation, it is important to ensure that the PPE is appropriate for the hazard. The same is true when protecting workers from exposure to azo dyes and the associated chemicals used in producing these dyes.

4.5 HUMAN BIOLOGICAL IMPACT

The biological impacts of azo dyes to humans are not completely known or understood. In fact, the knowledge in this area is minimal. Therefore, quantification of potential impacts on human offspring is not feasible. However, research has been performed in animals such as mouse, rats, hamsters, rabbits, and dogs[4] which show some impact on offspring from exposure to parent. More is known about the impact on the exposed individual and the potential development of cancer from dyes that have carcinogenic tendencies when metabolized in the body or exposed directly to a known carcinogen.

Some azo dyes and dye products can leave their scars on humans and can even cause death if the exposure that leads to the resultant disease is prolonged or severe enough. The impact of exposure can range from contact dermatitis to cancer and death.

4.6 TREATMENT OF RESULTANT ILLNESS

The treatment for exposure to dyes would be necessary when conditions exist for contact dermatitis in acute exposure, and the various types of cancers that can be caused by chronic exposure to these dyes. The treatment of contact dermatitis generally involves applying topical treatment such as a cream and protecting the skin from further exposure. On the other hand, the treatment for cancer is quite invasive and involves a long-term strategy with a team of physicians providing input into the type of medical care required. The type of medical care or treatment is dependent on the current overall health of the individual, the stage of cancer, the location in the body, and other determinants of care.

REFERENCES

1. Beyerbach, A., Rothman, N., Bhatnagar, V.K., Kashyap, R., & Sabbioni, G. (2006). Hemoglobin adducts in workers exposed to benzidine and azo dyes. *Carcinogenesis*, 27(8), 1600–1606.
2. Yamjala, K., Nainar, M.S., Ramisetti, N.R. (2015). Methods for the analysis of azo dyes employed in food industry – A review. *Food Chemistry*, 192, 813–821.
3. Yamiala, K., Nainar, M.S., Ramisetti, N.R. (2016). Method for the analysis of azo dyes employed in food industry – A review. *Food Chemistry*, 192, 813–821.
4. World Health Organization International Agency for Research on Cancer, IARC Monographs on the Evaluation of Carcinogenic Risks to Humans, 99, Some Aromatic Amines, Organic Dyes, and Related Exposures, Lyon France (2010).
5. Gao, Y., Li, C., Shen, J., Yin, H., An, X., & Jin, H. (2011). Effect of food azo dye tartrazine on learning and memory functions in mice and rats, and the possible mechanisms involved. *Journal of Food Science*, 76(6), T125–T129.
6. Chequer, F.M., Angeli, J.P., Ferraz, E.R., Tsuboy, M.S., Marcarini, J.C, Mantovani, M.S., de Oliveira, D.P. (2009). The zo yes disperse red 1 and disperse orange 1 increase the micronuclei frequencies in human lymphocytes and in HepG2 cells. *Mutation Research/Genetic Toxicology and Environmental Mutagenesis*, 676(1–3), 83–86.
7. Office of environmental health hazard assessment health effects assessment: potential neurobehavioral effects of synthetic food dyes in children public review draft (2020).
8. Malinauskiene, L., Bruce, M., Ryberg, K., Zimerson, E., Isakson, M. (2012). Contact allergy from disperse dyes in textiles–A review. *Contact Dermatitis*, 68, 65–75.
9. www.fsw.cc/azo-dye/.
10. Morrison, J.M., & John, G.H. (2015). Non-classical azoreductase secretion in clostridium perfringens in response to sulfonated azo dye exposure *Anerobe*, 34, 34–43.
11. Ryberg, K., Isaksson, M., Gruvberger, B., Hindsen, M., Zimerson, E., Bruze, M. (2006). Contact allergy to textile dyes in southern Sweden. *Contact Dermatitis*, 54, 313–321.
12. Gurdeep, R., Chatwal, & Madhu Arora (2008). Synthetic Dyes, Global Media,–12–01.
13. https//en.wikipedia.org/wiki/Category:Azo_dyes.
14. https//ww.fda.gov/industry/color-additives/color-additives-historyaddit.
15. Jiao, J., Wang, J., Li, M., Li, J., Li, Q., Quan, Q., & Chen, J. (2016). Simultaneous determination of three azo dyes in food product by ion mobility spectrometry. *Journal of Chromatography* B, 1025, 105-109.

Chapter 5

Carbamates

5.1 INTRODUCTION

Carbamate compounds are esters of carbamic acid that are usually used as insecticides. These compounds are referred to as N-methylcarbamates. Derivatives of carbamic acid, thiocarbamic acid, and dithiocarbamic acid are used as herbicides. They increase agricultural production, aside from protecting agricultural workers and animals from insect-vector-mediated diseases. Carbamates are a class of insecticides structurally and mechanistically similar to organophosphate (OP) insecticides. Carbamates are N-methyl carbamates derived from a carbamic acid and cause carbamylation of acetylcholinesterase (AChE) at neuronal synapses and neuromuscular junctions. While they possess a similar mechanism of action to the irreversible phosphorylation of AChE by OPs, carbamates bind to AChE reversibly. Carbamates cause reversible inhibition of the AChE enzyme, which is present at parasympathetic and sympathetic ganglia, parasympathetic muscarinic terminal junctions, sympathetic fibers located in sweat glands, and nicotinic receptors at the skeletal neuromuscular junction. Persistently elevated acetylcholine levels due to AChE inhibition lead to increased neurotransmitter signaling. Central nervous system symptoms from increased acetylcholine include confusion, delirium, hallucinations, tremor, and seizures. Increased acetylcholine levels in the autonomic nervous system increase sympathetic and parasympathetic activity. Classic mnemonics emphasize the parasympathetic symptoms from carbamate and OP toxicity. Toxicity is mediated through both cholinergic and noncholinergic modes. They are reported to carry out reversible carbamylation of active site amino acid residue serine of AChE, leading to accumulation of acetylcholine at synapses and neuromuscular junctions, causing a "cholinergic crisis." AChE, under normal conditions, catalyzes the hydrolysis of acetylcholine (neurotransmitter) to acetic acid and choline, leading to cessation of signaling. The carbamates are mainly used in agriculture, as insecticides, fungicides, herbicides, nematicides, or sprout inhibitors. In addition, they are used as biocides for industrial or other applications and in household products. Three classes of carbamate pesticides are known[1]:

DOI: 10.1201/9781003220114-5

- The carbamate ester derivatives, used as insecticides (and nemati-cides), are generally stable and have a low vapor pressure and low water solubility.
- The carbamate herbicides (and sprout inhibitors) have the general structure R1NHC(O)OR2, in which R1 and R2 are aromatic and/or aliphatic moieties.
- Carbamate fungicides contain a benzimidazole group.

There are more than 50 compounds classified as carbamates; these com-pounds are heavily used in agriculture as insecticides, fungicides, herbi-cides, nematicides and sprout inhibitors, biocides for industrial or other applications, and in household products for control of household pests. Carbamate insecticides are derivatives of carbamic acid and vary in their spectrum of activity, mammalian toxicity, and persistence and are used as either dusts or sprays.

5.1.1 Carbamate used as therapeutics

Carbamate has been increasingly gaining an important role in modern drug discovery and medicinal chemistry because of its very good chemical and proteolytic stability, ability to penetrate cell membranes, and resemblance to a peptide bond. Carbamates are an important structural component of drugs and prodrugs approved and marketed for the treatment of vari-ous diseases including cancer, epilepsy, hepatitis C, HIV infection, and Alzheimer's disease. In drugs, carbamates can play a role in drug-targeting or in improving the biological activity of parent molecules. When used in prodrugs carbamates are mainly used to delay first-pass metabolism and to enhance the bioavailability and effectiveness of compounds.

Some studies to date have shown that the carbamate group can increase the biological activity of active pharmacophores of some compounds. The carbamate group constitutes part of many approved drugs that act as che-motherapeutic agents, cholinesterase inhibitors in the treatment of neurode-generative disorders, physostigmine, human immunodeficiency virus (HIV) protease inhibitors, anticonvulsants, and muscle relaxants.[4] Table 5.1 lists some drugs and prodrugs where carbamates are an integral constituent.

5.2 EXPOSURE PATHWAYS

Toxic exposures to carbamates can occur via dermal, inhalational, and gas-trointestinal (GI) exposures. Symptom severity depends on the classification of the pesticide and the dose of the exposed individual.[4] Health hazards for people occur mainly from occupational over-exposure to carbamate insecticides resulting in poisoning characterized by cholinergic symptoms caused by inhibition of the enzyme AChE. Various cases of intoxication

Table 5.1 Carbamate medicinal benefits

Drug	Carbamate role
Docetaxel	Prolongs drug action, increases drug potency, improves water solubility
Mitomycin C	Participates in the formation of an alkylating compound during reaction with the target
Rivastigmine, neostigmine, physostigmine, pyridostigmine	Key element for interaction with the target
Ritonavir, amprenavir, atazanavir, darunavir	Improves drug bioavailability and potency, engaged in a backbone interaction with protease
Ombitasvir, elbasvir, daclatasavir	Improves drug stability and lipophilicity
Febendazole, mebendazole, febantel, albendazole	Improves aqueous solubility and bioavailability, increases cytotoxicity
Mehocarabamol, metaxalone	Inhibits acetylcholinesterase at synapses in the autonomic nervous system, neuromuscular junction, and central nervous system
Felbamate	Improves drug stability and bioavailability
Retigabine	Major pharmacophore responsible for interacting with residues in the KCNQ2–5 channels
Gabapentin enacarbil	Improves bioavailability
Capecitabine	Improves selectivity and bioavailability
Bambuterol	Delays first-pass metabolism
Irinotecan	Improves aqueous solubility

have been recorded, and most of them were workers applying insecticides inside houses in the tropics to control mosquito vectors of malaria, or plant protection workers. The main routes of exposure are inhalation and skin. Certain carbamates may reach groundwater and consequently may find their way into drinking water. Some of the common ways workers handling carbamate pesticides are exposed are shown in Table 5.2[2]:

5.2.1 Inhalation exposures and impact

Carbamates are absorbed by inhalation, ingestion, and through the skin, although exposure through the skin tends to be the less-toxic route. Exposure can result from combined dermal and inhalational exposures after working in areas recently sprayed or fogged with insecticides. Pediatric cases caused by playing on a sporting field after insecticide spraying have been reported. After massive exposures, patients may become symptomatic within 5 minutes. The time to symptom onset is dependent on the exposure dose and the toxicity of the given carbamate. Highly lipophilic carbamates will redistribute into fat stores from the extracellular fluid quickly and have decreased clinical effects initially. Carbamates are hepatically metabolized via hydrolysis, hydroxylation, and conjugation, and 90% are renally excreted in a

Table 5.2 Exposure route/pathway

Exposure route	Exposure pathway
Dermal	• Not washing hands after handling pesticides, containers, or equipment
	• Splashing/spilling pesticide on skin
	• Not wearing gloves when removing pesticide- contaminated personal protective equipment
	• Applying pesticides in windy weather
	• Not wearing gloves when touching treated plants or when handling spray equipment
Oral	• Not washing hands before eating, smoking, or chewing gum
	• Splashing pesticide into mouth
	• Storing pesticide in anything but the original container
Inhalation	• Handling pesticides in confined or poorly ventilated areas without wearing a respirator
	• Handling dust or powders without wearing a respirator
	• Using an inadequate or poorly fitting respirator
	• Being exposed to drift without wearing a respirator
	• Not washing hands before smoking
Eye	• Rubbing eyes or forehead with contaminated gloves or hands
	• Splashing pesticide in eyes
	• Pouring dry formulations without wearing goggles
	• Applying pesticides in windy weather without wearing goggles

matter of days. Data are conflicting on CNS and cerebrospinal fluid penetration of carbamates. Adults tend to have less CNS toxicity, whereas, in pediatric exposures, CNS depression is often a predominant symptom. Importantly, carbamates do not undergo the "aging" that occurs during the phosphorylation of OPs to AChE, and the carbamate-cholinesterase bond hydrolyzes spontaneously within hours.[2]

5.2.2 Dermal exposures and impact

Acute dermal toxicity values for most carbamates are mainly above 500 mg/kg body weight except for aldicarb, which is highly toxic.[1] Dermal absorption appears to be low with increasing absorption in cases of disruption in the skin and exposure to highly toxic carbamates. Rat data show peak inhibition of cholinesterases by 30 minutes after oral administration[2].

5.3 EXPOSURE CONTROL METHODS

Work involving carbamates should be performed as an enclosed operation and use of local exhaust ventilation at the site of chemical release where possible. Exposure of workers handling insecticides to carbamates should be prevented by wearing personal protective equipment (PPE). Neoprene or nitrile gloves provide adequate protection from cutaneous exposures, and the worker should wear full PPE with a minimum of a gown, respirator, and face shield. Latex gloves do not provide adequate protection from insecticides. Workers exposed to carbamates should be decontaminated because of cutaneous absorption of carbamate pesticides. Decontamination should take place as soon as possible. All clothing worn by the worker should be removed, and the skin should be triple-washed with water, then soap and water, and then rinsed again with water.

5.3.1 Product elimination and substation

Of the total carbamates applied as pesticides in the agriculture sector, only a minor fraction impacts the target organisms, whereas the rest of it is distributed into the environment, harming non-target biota, and leading to ecological imbalance. Processes that mainly contribute to the distribution and persistence of these compounds include leaching, adsorption, run-off, volatilization, and partial degradation (by both biotic and abiotic factors)

5.3.2 Engineering controls

The best option for protecting workers against carbamates is to design safety into the process and systems. Engineering control when effective will eliminate or minimize worker exposure to carbamates thereby minimizing exposure potential. Engineering controls to assist with controlling exposures to carbamates if used in the laboratory include laboratory hoods, ventilated enclosures, local exhaust ventilation system, and tight-fitting lids or fixtures. When using these control methods, a maintenance program must be developed and implemented to ensure that these measures are and remain effective for its intended purpose and usage as a safety control measure.

5.3.3 Administrative controls

Administrative controls should be implemented by employers to control exposures to the different types of carbamates used in the workplace. It is used with other controls. It is easier to control exposures administratively when used in the work environment. Consumers using these chemicals at home are not likely to automatically think of the potential hazards and seek to implement controls beyond the ones recommended on the label. Therefore, from an administrative control perspective, manufacturers must include as much information on ways consumers can protect themselves.

5.3.4 Protective equipment and effectiveness

PPE is used commonly by workers that use carbamates to control exposures to carbamates. Gloves are used by workers to protect against skin exposure when handling carbamates. When selecting a glove to protect against skin exposure consult a glove chart or manufacturer to ensure the selection of the appropriate glove. Use safety glasses to prevent carbamate exposure to the eyes. Use proper protective clothing. Do not breathe dust or spray mist.

In the work environment because of the diversity of hazards that may be involved in any one task or process, it is recommended that an evaluation be conducted to ensure that all PPE needs are identified and the appropriate PPE is selected. More information on the use and selection of PPE is found in Appendix B.

5.4 HUMAN BIOLOGICAL IMPACT

The impact on humans with regard to exposure to carbamates depends on the effects of the metabolites on target organs. Apart from hazards emerging from cholinergic excess, carbamate pesticides and their metabolites act as endocrine disruptors by potentiating or antagonizing the activity of steroid hormones, thus disrupting steroidogenesis and thyroid function. Carbamate pesticides (and their metabolites) mimic various hormones and bind to their receptors, therefore affecting the expression of the responsive genes. The estrogen receptors α and β (ERα and ERβ) are estrogen-binding transcription factors that regulate steroid hormone-mediated gene expression.

5.4.1 Potential impact on exposed individual

Exposure may be chronic or acute and absorbed from the skin, lungs, conjunctiva, mucous membranes, lungs, and GI tract. Dermal absorption appears to be low with increasing absorption in cases of disruption in the skin and exposure to highly toxic carbamates. Rat data show peak inhibition of cholinesterases by 30 minutes after oral administration.

After massive exposures, patients may become symptomatic within 5 minutes. The time to symptom onset is dependent on the exposure dose and the toxicity of the given carbamate. Highly lipophilic carbamates will redistribute into fat stores from the extracellular fluid quickly and have decreased clinical effects initially.

Carbamates are hepatically metabolized via hydrolysis, hydroxylation, and conjugation, and 90% are renally excreted in a matter of days. Data are conflicting on CNS and cerebrospinal fluid penetration of carbamates. Adults tend to have less CNS toxicity, whereas in pediatric exposures, CNS depression is often a predominant symptom. Importantly, carbamates do not undergo the "aging" that occurs during the phosphorylation of OPs to AChE, and the carbamate-cholinesterase bond hydrolyzes spontaneously within hours. Carbamate exposure can cause headaches, dizziness, or weakness. It can make you feel like you will throw up. It can also cause shaking, stomach cramps, diarrhea, and sweating. Skin exposure to carbamates causes a minor rash. Long-term exposure can result in loss of appetite, weakness, weight loss, and a general feeling of sickness. Ethyl carbamate (urethane) is a well-known carcinogen, and it seems that its chemical structure is optimal for such an effect. Any change in the molecule seems to decrease the carcinogenic potency, particularly when the ethyl group is replaced by larger side chains. Alkyl groups on the nitrogen also reduce this activity. The carcinogenicity studies with benzimidazole derivatives showed either positive or equivocal results. Carbamate pesticides may be converted to N-nitroso compounds. These N-nitroso compounds be considered mutagenic and carcinogenic. However, the amount of nitroso compounds that can be expected to result from dietary intake of carbamate pesticide residues is negligible in comparison with nitroso-precursors that occur naturally in food and drinking water.

5.4.2 Potential impact on off-spring

The use of pesticides to reduce the mosquito vector populations is the foundation used across the globe to control malaria control. The biological impact of most pesticides on human populations, including pregnant women and infants, is not completely known. However, some pesticides, including carbamates, have been shown to distress the human immune system. A study was conducted on the effects of bendiocarb that is a commonly used carbamate to determine potential health impacts. The systemic absorption and immunologic effects of bendiocarb following household spraying in a cohort of pregnant Ugandan women and their infants were measured. The researchers concluded that bendiocarb was present at high levels in maternal, umbilical cord, and infant plasma of individuals exposed during pregnancy. These researchers also concluded from their study that bendiocarb is systemically absorbed and trans-placentally transferred to the fetus and bendiocarb exposure is associated with many changes in fetal immune cell homeostasis and function[7]. Other studies have linked carbamate pesticides to fetal growth issues. Therefore, considerations must be given to avoiding potential exposure to pregnant workers. To avoid exposures to pregnant women the following steps should be taken.

- Ensure that a detailed hazard and risk analysis of the area where carbonate products are used, manufactured, handled, and stored is completed and the information contained in the analysis is shared with all workers.
- Provide training to workers on the hazards associated with carbonates.
- As soon as a pregnancy is declared, the employee should be removed from the work area where a potential exposure exits.

5.5 TREATMENT OF RESULTANT ILLNESS

The diagnosis and management of carbamate toxicity should be made by an interprofessional medical team that includes the emergency department physician, pulmonologist, neurologist, nurse practitioner, emergency department nurse, intensivist, and pharmacist. Attempts should be made to gain a history of potential known exposures making use of information that can be found on the product MSDS. Ingestion of small amounts of carbamates that result in mild cases, a full recovery is generally expected; however, severe cases can lead to long-term hospitalization accompanied with difficulty in breathing[2].

Carbamate poisoning is treated with a substance called atropine. When dermal exposure occurs, decontamination procedures include the removal of contaminated clothes and washing of the skin with alkaline soap or with a sodium bicarbonate solution. Care should be taken in the cleaning of the skin area where venipuncture is performed. Blood might be contaminated with carbamates and, therefore, inaccurate measures of ChE inhibition might result. Extensive eye irrigation with water or saline should also be performed. In the case of ingestion, vomiting can be induced, if the patient is conscious, by the administration of ipecacuanha syrup (10–30 ml) followed by 200 ml of water. However, this treatment is contraindicated in the case of pesticides dissolved in hydrocarbon solvents. Gastric lavage (with the addition of bicarbonate solution or activated charcoal) can also be performed, particularly in unconscious patients, taking care to prevent the aspiration of fluids into the lungs (i.e., only after a tracheal tube has been put in place).

Decontamination

When there is a continued cutaneous absorption of carbamate pesticides, decontamination should take place as soon as possible. Medical providers should avoid self-contamination by wearing PPE. Neoprene or nitrile gloves provide adequate protection from cutaneous exposures, and the provider should wear full PPE with a minimum of a gown, mask, and face shield. Latex gloves do not provide adequate protection from insecticides. All clothing should be removed from the patients, and the skin should be triple-washed with water, then soap and water, and then rinsed again with water. Vomitus and diarrhea may cause cutaneous absorption in providers in cases of GI ingestions. In massive, life-threatening ingestions, GI decontamination may be considered if

- the patient has not had bouts of emesis
- the ingestion occurred within 1 hour
- if the patient is protecting their airway.

In this instance, nasogastric lavage can be instituted. In severe toxicity, patients may have seizures, respiratory paralysis, and coma. Airway protection should take place before GI decontamination if any of these features are present. Data are disputed regarding carbamate toxicity's adequate adsorption by activated charcoal. Some experts recommend administering 1 g/kg of single-dose activated charcoal if the patient presents within 1 hour of a massive life-threatening GI ingestion. Consultation with the poison center or regional toxicologist before GI decontamination may be a reasonable approach given the risk of aspiration of activated charcoal and the questionable benefit of this therapy.

Respiratory

Respiratory failure and hypoxemia are the primary causes of death after toxic exposure to AChE inhibitors. This is multifactorial secondary to bronchorrhea, muscular weakness with potential flaccid paralysis, and depression of CNS respiratory drive. After decontamination, initial patient assessment should be directed at ensuring adequate ventilation and oxygenation. Increased respiratory secretions may be treated with atropine via competitive inhibition of the excessive muscarinic receptor excitation. Early endotracheal intubation should be performed for patients with difficulty managing their respiratory secretions, comatose or severely depressed mental status, or significant skeletal muscle weakness. Depolarizing neuromuscular blockers such as succinylcholine should be avoided, as serum cholinesterases are inactivated by AChE inhibitors, and prolonged paralysis lasting up to several hours can occur. Instead, paralysis should be induced using nondepolarizing neuromuscular blockers such as rocuronium.

Atropine

Atropine competitively antagonizes the increased acetylcholine levels at muscarinic receptors and decreases the symptoms of lacrimation, salivation, miosis, emesis, diarrhea, diaphoresis, urinary incontinence, bronchospasm, and excessive respiratory secretions. Atropine, starting at doses of 1 – 3 mg intravenously (IV) in adults or 0.05 mg/kg IV in pediatric patients with a minimum dose of 0.1 mg, should be administered. The dose should be doubled every 5 minutes if the previous dose provides an inadequate response. Previous descriptions of "atropinization" (dry skin and mucous membranes, decreased bowel sounds, tachycardia, an absence of bronchospasm, and mydriasis) did not emphasize meaningful endpoints of resuscitation and treatment should be directed toward achieving cardiorespiratory stability. An adequate dose of atropine is reached when there is attenuation of tracheobronchial secretions and decreasing bronchoconstriction accompanied by adequate blood pressure and heart rate for tissue perfusion. After a stabilizing dose of atropine is reached, treatment response is maintained by a constant infusion of atropine that is usually 10% – 20% of the bolus dose per hour. Tachycardia is not a contraindication to atropine administration in patients presenting with carbamate poisoning, as tachycardia may be secondary to hypoxia and excessive bronchopulmonary secretions. Doses over 1000 mg of atropine have been recorded over 24 hours to treat severe AChE inhibitor poisonings. Atropine does not reverse the skeletal muscle weakness caused by nicotinic receptor stimulation in carbamate toxicity. Patients need to have continued monitoring for potential respiratory failure, requiring mechanical ventilation after atropine administration.

Oxime

Pralidoxime (2-PAM) is commonly given to patients with OP toxicity early in the presentation to prevent the "aging" process as OPs irreversibly bind to AChE. Carbamates will spontaneously disassociate from AChE and recover function within 24 – 48 hours. Studies have shown potentially increased AChE inactivation if pralidoxime is administered in cases of carbaryl poisoning. However, the potential benefit of oxime therapy in aldicarb poisoning has been described. In cases of known single-agent carbamate toxicity without concern for possible concomitant OP exposure, pralidoxime therapy can be withheld. However, when faced with undifferentiated insecticide toxicity, pralidoxime can be given, as administration in carbamate toxicity is unlikely to be detrimental, and the benefit for OP intoxication is well described.

Benzodiazepines

Benzodiazepines are used for the treatment of seizures and agitation in intubated patients after carbamate toxicity. Limited data exist evaluating the efficacy of benzodiazepines for seizures secondary to insecticide poisonings, as seizures are uncommon in large case series of carbamate and OP toxicity. Due to this lack of data, standard abortive seizure therapy with benzodiazepines is commonly instituted.

Disposition

Carbamates typically have a more benign clinical course compared to OP poisonings due to transient cholinesterase inhibition and rapid reactivation of AChE enzymatic activity. Most patients will experience complete recovery within 24 hours. Patients who have depressed levels of consciousness can have significant mortality. Patients with mild initial symptoms not requiring atropine can be safely discharged after observation. Moderate

Table 5.3 Relationship of chemical structure and pesticidal activity of carbamates[5]

Pesticidal activity	Other names	Chemical structure
Herbicide	Barban, chlorbufam, desmedipham, phenmedipham, swep, carbetamide, dichlormate, asulam, karbutilate, terbucarb	
Herbicide and sprout inhibitors	Chlorpropham,	
	Propham	
Fungicide	Benomyl, thiophanate-methyl, thiophanate ethyl, carbendazim	
Insecticide	Aldoxycarb, aminocarb, BPMC, bendiocarb, butacarb, carbanolate, carbaryl, bufencarb, carbofuran, cloethocarb, dimetilan, methiocarb aldicarb	

Table 5.4 Carbamate insecticide properties

Name, molecular weight, property	Chemical formula	Acute oral LD50 (rat, mg/kg)
Bendiocarb MW: 223.2 MP: 125–129	$C_{11}H_{13}N_{04}$	40~156
Benfuracarb MW: 410.5 MP: 110	$C_{20}H_{30}N_2O_5S$	205
Carbaryl MW: 201.2 MP: 142	$C_{12}H_{15}NO_2$	500
Carbofuran MW: 221.3 MP: 153–154	$C_{12}H_{15}NO_3$	8
Carbosulfan MW: 380.5 BP: 124–128	$C_{20}H_{32}N_3O_3S$	623
Ethiofencarb MW:: 225.3 MP: 33.4	$C_{11}H_{15}NO_2S$	200
Fenobucarb MW: 207.3 MP: 31–32	$C_{12}H_{17}NO_2$	185
Furathiocarb MW: 382.5 BP: 250<	$C_{18}H_{26}N2O_5S$	53
Isoprocarb MW: 225.3 MP: 93–96	$C_{11}H_{15}NO_2$	450
Metolcarb MW: 162.2 MP: 76 –77	$C_9H_{11}NO_2$	498
Methomyl MW: 162.2 MP: 78–79	$C_5H_{13}N_2O_2S$	17
Oxamyl MW: 219.3 MP: 100–102	$C_7H_{13}N_3O_3S$	5.4
Pirimicarb MW: 238.3 MP: 90.5	$C_{11}H_{18}N_4O_2$	147

(Continued)

Table 5.4 (Continued) Carbamate insecticide properties

Name, molecular weight, property	Chemical formula	Acute oral LD50 (rat, mg/kg)
Propoxur MW: 209.2 MP: 90	$C_{11}H_{15}NO_3$	50
Thiodicarb MW: 354.5 MP: 173–174	$C_{10}H_{18}N_4O_4S_3$	66
XMC MW: 179.2 MP: 99	$C_{10}H_{13}NO_2$	542
Xylylcarb MW: 179.2 MP: 79–80	$C_{10}H_{13}NO_2$	325

MF, molecular formula; MW, molecular weight; MP (°C), melting point; BP (°C), boiling point

poisonings will necessitate 24 hours of observation, and patients requiring atropine should be admitted to a monitored setting for continued assessment of their respiratory status.

5.6 TECHNICAL INFORMATION

There are at least 40 carbamate insecticide numbers. Therefore, the CAS numbers are too numerous to provide. Chemical Formula: There are at least 40 carbamate insecticides. Therefore, the chemical formulas are too numerous to provide. Carcinogenicity (EPA): A review of the EPA/IRIS database found that the carbamate insecticides listed were either not assessed under the program or not classifiable as to their carcinogenicity. MCL (drinking water): The MCL for carbofuran is 0.04 mg/L. Aldicarb is not listed on the EPA List of Drinking Water Contaminants and MCLs; however, an EPA document references a proposed MCL for Aldicarb of 3 ppb. OSHA standards: There is an OSHA PEL standard carbaryl of 5 mg/m³. NIOSH standards: There is a NIOSH REL time-weighted average for propoxur of 0.5 mg/m³. The NIOSH REL for carbofuran is 0.1 mg/m³. The NIOSH REL standard for carbaryl is 5 mg/m³. The NIOSH limit for methomyl is 2.5 mg/m³ averaged over a 10-hour

work shift. Since there are over 40 carbamate insecticides, this list should not be considered inclusive of the occupational exposure limits[6].

Table 5.3 shows the relationship of between chemical structure and pesticidal activity of carbamates,[5] and Table 5.4 shows carbamate insecticides with their chemical formulas, weight, and toxicities.

REFERENCES

1. World Health Organization. (1986). Carbamate pesticides: a general introduction. World Health Organization. https://apps.who.int/iris/bitstream/handle/10665/38687/9241542640-eng.pdf.
2. Silberman, J., & Taylor, A. (2018). Carbamate toxicity. StatPearls Publishing, Treasure Island (FL). PMID: 29489157.
3. Matošević, A., & Bosak, A. (2020). Carbamate group as structural motif in drugs: a review of carbamate derivatives used as therapeutic agents. *Arhiv za higijenu rada i toksikologiju, 71*(4), 285–299.
4. Government of Newfoundland and Labrador (2022). Human health, landscape. *In gov.nl.ca*. Retrieved From https://www.google.com/url?sa=t&rct=j&q=&esrc=s&source=web&cd=&ved=2ahUKEwjXlte27sn4AhWsKkQIHdbTBtkQFnoECAsQAQ&url=https%3A%2F%2Fwww.gov.nl.ca%2F&usg=AOvVaw34wzmvvFrjsJT1mSpAN9R- https://www.gov.nl.ca/ecc/files/env-protection-pesticides-business-manuals-landscape-chapter2.pdf
5. Mdeni, N.L., Adeniji, A.O., Okoh, A.I., & Okoh, O.O. (2022). Analytical evaluation of carbamate and organophosphate pesticides in human and environmental matrices: a review. *Molecules: a Journal of Synthetic Chemistry and Natural Product Chemistry, 27*(3), 618.
6. University of Wisconsin-Madison (2022). Occupational health-other pesticides. Wisconsin Occupational Health Laboratory, University of Wisconsin-Madison. http://www.slh.wisc.edu/occupational/wohl/analytical-services/pesticides/other-pesticides/

Chapter 6

Halogenated compounds

6.1 INTRODUCTION

The chemicals that are known as halogens are listed in group 17 of the periodic table. These elements are fluorine (F), chlorine (Cl), bromine (Br), iodine (I), and astatine (At). Together, these elements form what is commonly known as the halogen group. Astatine has no natural or synthetic isotopes that can be considered stable[2,3] and will not be of focus in this chapter. Fluorine (F_2) has the highest crustal abundance of the halogens (544 mg/kg) while iodine has the lowest (0.25 mg/kg). Chlorine is the most abundant halogen in the universe. The chemical composition of the four naturally occurring halogens has some similarities with fluorine, chlorine, and bromine being classified as lithophile elements while iodine has a more chalcophile nature. Iodine and chlorine are essential elements for mammals, and fluorine has been shown to have beneficial effects on bone and tooth formation.[4,5]

Fluorine is the most electronegative and most reactive of all the periodic elements. It can form binary compounds with all other elements apart from helium, neon, and argon. It is a pale-yellow, highly reactive gas and a strong oxidant. Fluorine is a toxic irritant that attacks the skin and the mucous membranes of the exposed individual's nose and eyes.[7] Fluorine chemistries have important industrial uses. For example, preparation of Freons in the 1930s was a major advance in the refrigeration field. F_2 has also increasingly become an important laboratory reagent for the synthesis of some fluorinated molecules, many having important biological and medical applications.[3]

Chlorine gas is produced commercially by electrolysis of sodium chloride (NaCl) brine in asbestos diaphragm cells or mercury cathode cells. Of all the halogens, Cl_2 has the most expansive industrial use and is ranked high among the large volume of chemicals manufactured in the United States in 1979. Some primary uses include the production of organic compounds; use in bleaches for paper, pulp, and textile, sanitation, and disinfection of water supplies; and production of inorganic chemicals. Chlorine gas (Cl_2) reacts with water to produce strong oxidizing solution. This reaction can damage the moist tissue lining of the respiratory tract when exposed to chlorine.[7]

DOI: 10.1201/9781003220114-6

Bromine is prepared commercially by the oxidation of bromide with Cl_2. The production of 1,2-dibromoethane as a scavenger for lead in gasoline had dominated the industrial uses of Br_2 in previous years. The use of bromine-containing compounds as pesticides and fire retardants is an important industrial application of Br_2. Bromine is a highly corrosive liquid having vapor that is extremely irritating to the eyes, skin, and mucous membranes. Short exposure to $40 - 60$ ppm can pose a serious hazard to the exposed individual, and exposure at 1,000 ppm is fatal. While inhalation of small amounts of bromine can lead to coughing, nosebleed, a feeling of oppression, vertigo, headache, abdominal pain, diarrhea, and asthma-like respiratory issues, sometimes accompanied by measles-like lesions on the skin[3]. Bromine (Br_2) is a toxic, volatile, dark-red liquid when inhaled or ingested. It is strongly irritating to the mucous tissue, eyes, and respiratory tract and may cause pulmonary edema. The irritating odor of bromine causes a withdrawal action from the point of exposure; therefore, it is not easy to collect toxicological data.[7]

Iodine is a purple-to-black crystalline solid with a distinctive strong odor. It is used in industries such as pharmaceuticals, dye photographic material, lithography, antiseptic, and X-ray contrast. Exposure to iodine can irritate the throat and burn eyes and skin. Exposure to iodine can lead to skin allergy, headache, vomiting, diarrhea, abdominal pain, or thyroid disturbances.[12] Iodine (I_2) is a solid that is irritating to the lungs. The relatively low vapor pressure of iodine limits exposure potential to the vapor.[7] The prominent halogens and basic identifying information are listed in Table 6.1.

6.2 HALOGENATED HYDROCARBONS

Halogenated hydrocarbons, sometimes referred to as halocarbons, are hydrocarbon compounds that have at least one hydrogen atom which is replaced by a halogen atom. Halogens are found in Group VIIA of the periodic table (Table 6.1). These atoms are included in Table 6.1. Common examples of halogenated hydrocarbons are listed in Table 6.2.

Table 6.1 Prominent halogens

Halogen	Trade name	CAS no.	Physical description
Fluorine	Flourine-19	7782-41-4	Pale yellow to greenish gas with a pungent irritating odor
Chlorine	Molecular chlorine	7782-50-5	Greenish-yellow gas with a pungent irritating odor
Bromine	Molecular bromine	7726-95-6	Dark reddish-brown fuming liquid with suffocating irritating fumes
Iodine	Iodine crystals, molecular iodine	7553-56-2	Violet solid with a sharp characteristic odor

Table 6.2 Common halogenated hydrocarbons

Halogenated hydrocarbon	Trade names/synonyms	CAS no.
1-Bromopropane	N-Propyl bromide, propyl bromide, propane	111-30-8
Methylene chloride	Dichloromethane	75-09-2
Chloroform	Methane trichloride, trichloromethane	67-66-3
Tetrachloroethylene	Perchloroethylene, perk, tetrachloroethylene, perchloroethylene	127-18-4
Carbon tetrachloride	Carbon chloride, carbon tetrachloride, halon-104, tetrachloroethane, refrigerant 10	56-23-5

Halogenated hydrocarbons have a wide variety of uses.

- Tetrachloroethylene, also commonly referred to as perchloroethylene, has been extensively used as a solvent in the dry-cleaning industry.
- Methylene chloride has been effectively used as a degreasing solvent for cleaning metals and for the spray application of adhesives.
- In recent years, 1-bromopropane has been introduced as a substitute for tetrachloroethylene in dry-cleaning and for methylene chloride in spray adhesives.
- Chlorofluorocarbons (Freon) have been used as refrigerants.
- Chlorinated hydrocarbons have been used as insecticides and herbicides.[1]

6.3 HALOGENATED ORGANIC COMPOUNDS

Halogenated organic compounds represent a large class of natural and synthetic chemicals that contain one or more halogens (fluorine, chlorine, bromine, or iodine) combined with carbon and other elements. The simplest organochlorine compound is chloromethane, also called methyl chloride (CH_3Cl) (Organic Halogen Compounds – Chemistry Encyclopedia – reaction, elements, examples, gas, number, atom, synthesis, reactivity (chemistryexplained.com).

Sodium hypochlorite (bleach – NaOCl) and many organic chemicals contained in household cleaning products can react to produce halogenated volatile organic compounds (VOCs) as validated through experimentation.[6] The experiment results also add credence to the fact that bleach can be important in terms of inhalation of carbon tetrachloride, chloroform, and several other halogenated VOCs.[6] With this knowledge, consumers must keep in mind that usage of halogenated organic compounds comes with risks. These risks can be minimized by knowing the product and using the products as prescribed while having in place the safety requirement during usage. Some halogenated organic compounds are listed in Table 6.3.

Table 6.3 Some halogenated organic compounds

Bromodichloromethane	p-Chloroaniline
Bromomethane	Chlorobenzilate
Carbon tetrachloride	p-Chloro-m-cresol
Chlorobenzene	2-Chloronaphthalene
2-Chloro-1,3-butadiene	2-Chlorphenol
Chlorodibromomethane	3-Chloropropionitrile
Chloroethane	m-Dichlorobenzene
2-Chloroethyl vinyl ether	o-Dichlorobenzene
Chloroform	p-Dichlorobenzene
Chloromethane	3.3'-Dichlorobenzidine
3-Chloropropene	2,4-Dichlorophenol
1,2-Dibromo-3-chloropropane	2,6-Dichlorophenol
1,2-Dibromomethane	Hexachlorobenzene
Dibromomethane	Hexachlorobutadiene
Trans-1,4-Dichloro-2-butene	Hexachlorocyclopentadiene
Dichlorodifluoromethane	Hexachloroethane
1,1-Dichloroethane1	Hexachloroprophene
1,2-Dichloroethane	Hexachlorpropene
1,1-Dichloroethylene	4,4'-Methylenebis(2-chloroanaline)
Trans-1,2-Dichloroethene	Pentachlorobenzene
1,2-Dichloropropane	Pentachloroethane
Trans-1,3-Dichloropropene	Pentachloronitrobenzene
cis-1,3-Dichloropropene	Pentachloronitrobenzene
Iodomethane	Pentachlorophenol
Methylene chloride	Pronamide
1,1,1,2-Tetrachloroethane	1,2,4,5-Tetrachlorobenzene
1,1,2,2-Tetrachloroethane	2,3,4,6-Tetrachlorophenol
Tetrachloroethene	1,2,4-Trichlorobenzene31.
1,1,1-Trichloroethane	2,4,5-Trichlorophenol
1,1,2-Trichloroethane	2,4,6-Trichlorophenol
Trichlorothene	Tris(2,3-dibromopropyl)phosphate
Trichloromonofluoromethane	Aldrin
1,2,3-Thrichloropropane	alpha-BHC
Vinyl chloride	beta-BHC
Bis(2-chloroethoxy)ethane	delta-BHC
Bis(2-chloroethyl)ether	gamma-BHC
Bis(2-chloroisopropyl)ether	Chlorodane
DDD	Aroclor 1232
DDE	Aroclor 1242
DDT	Aroclor 1248
Dieldrin	Aroclor 1254
Endosulfan I	Aroclor 1260

(*Continued*)

Table 6.3 (Continued) Some halogenated organic compounds

Endosulfan II	Hexachlorodibenzo-p-dioxins
Endrin	Hexachlorodibenzofuran
Endrin aldehyde	Pentachlorodibenzo-p-dioxins
Heptachlor	Pentachlorodibenzofuran
Heptachlor epoxide	Tetrachlorodibenzo-p-dioxins
Isodrin	Tetrachlorodibenzofuran
Kepone	2,3,7,8-Tetrachlorodibenzo-p-dioxin
Methoxyclor	
Toxaphene	
2,4-Dichlorophenoxyacetic acid	
Silvex	
2,4,5-T	
Aroclor 1016	
Aroclor 1221	

6.4 HALOGENATED SOLVENTS

Halogenated solvents are volatile organic chemicals containing a hydrocarbon chain or one hydrocarbon substituted with one or more chlorine or bromine atoms. Many of these chemicals are used as degreasers and solvents in a diversity of products such as paints. Some past and current usages include the following:

- 1,1,1-Trichloroethane was previously used as a dry-cleaning agent, insect fumigant, and solvent in some consumer products.
- Production of 1,1,2,2-tetrachloroethane in the United States has stopped, and currently, it is used as a chemical intermediate in the production of several other halogenated solvents.
- 1,1,2-Trichloroethane, 1,2-dichloroethane, and 1,1-dichloroethene are used in the synthesis of other chemicals, such as polyvinylidene.
- Tetrachloromethane use as a solvent and fumigant has been discontinued because of toxicity concerns and its other major use in the production of chlorofluorocarbon refrigerants, has been restricted because of ozone-depleting chemicals regulations.

Halogenated solvents may be released into the air from facilities that produce or use them, such as from contaminated wastewater, or from hazardous waste sites. These halogenated solvents generally do not persist in soil or water because of their volatility. Workers involved in the production or use of these solvents may be exposed primarily through inhalation or dermal contact with the liquid solvents. Some details about halogenated solvents are provided in Table 6.4.

Table 6.4 Halogenated solvents

Solvent	Synonym	CAS no.
Benzyl chloride	Chloromethylbenzene	100-44-1
Bis (2-chloroethyl)ether	Bis(-chloroethyl)ether	111-44-4
Bis(2-chloroisopropyl)ether	Bis(-chloroisopropyl)ether	108-60-1
Bromobenzene	Phenyl bromide	108-86-1
Bromochloromethane	Chlorobromomethane	74-97-5
Bromoethane	Ethyl bromide	74-96-4
Bromoform	Tribromomethane	75-25-2
Carbon tetrachloride	Tetrachloromethane	56-23-5
Chlorobenzene	Benzene chloride	108-90-7
2-Chloroethyl vinyl ether	(2-Chlorethoxy)ethane	110-75-8
Chloroform	Trichloromethane	67-66-3
1-Chloro-1-nitropropane	Chloronitropropane	600-25-9
2-Chlorophenol	o-Chlorophenol	95-57-8
4-Chlorophenyl phenyl ether	p-Chlorodiphenyl ether	7005-72-3
Chloropicrin	Trichloronitromethane	76-06-2
m-Chlorotoluene	NA	108-41-8
o-Chlorotoluene	2-Chloro-1-methylbenzene	95-45-8
p-Chlorotoluene	NA	106-43-4
Dibromochloromethane	Chlorodibromomethane	124-48-1
1,2-Dibromo-3-chlorpropane	DPCP	96-12-8
Dibromodifluoromethane	Freon 12-B2	75-61-6
1,2-Dichlorobenzene	o-Dichlorobenzene	95-50-1
1,3-Dichlorobenzene	m-Dichlorobenzene	541-73-1
1,1-Dichloroethane	1,1-DCA	75-34-3
1,2-Dichloroethane	Ethylene dichloride; 1,2-DCA	107-06-2
1,1-Dichloroethene	Vinylidene chloride; 1,1-DCE	75-35-4
trans-1,2-Dichloroethene	trans-1.2-DCE	156-60-5
1,2-Dichloropropane	Propylene dichloride	78-87-5
cis-1,3-Dichloropropene	cis-1,3-Dichloropropylene	10061-01-5
trans-1,3-Dichloropropene	trans-1,3-Dichlonopropylene	10061-02-0
Ethylene dibromide	1,2-Dibromoethane; EDB	106-93-4
Hexachlorobutadiene	HCBD	87-68-3
Hexachlorocyclopentadiene	HCCPD	77-47-4
Methylene chloride	Dichloromethane	75-09-2
Pentachloroethane	Ethane pentachloride	76-01-7
1,1,2,2-Tetrabromoethane	Acetylene tetrabromide	79-27-6
1,1,2,2-Tetrachloroethane	Acetylene tetrachloride	79-34-5
Trichloroethene	TCE	79-01-6
1,1,2-Trichlorofluoromethane	Freon 11	75-69-4
1,2,3-Trichloropropane	Allyl trichloride	96-18-4
1,1,2-Trichlorotrifluoroethane	Freon 113	76-13-1

(Continued)

Table 6.4 (Continued) Halogenated solvents

Solvent	Synonym	CAS no.
Carbon tetrachloride	Tetrachloromethane	56-23-5
Chloroform	Trichloromethane	67-66-3
1,1,1-Trichloroethane	Methyl chloroform; 1,1,1-TCA	71-55-6
Bromoethane	Ethyl bromide	74-96-4
Bromochloromehtane	Chlorobromomethane	74-97-5
Methylene chloride	Dichloromethane	75-09-2
Bromodichloromethane	Dichlorobromomethane	75-27-4
Bromoform	Tribromomethane	75-25-2
1,1-Dichloroethane	1,1-DCA	75-34-3
1,1-Dichloroethene	Vinylidene chloride; 1,1-DCE	75-35-4
Dibromodifluoromethane	Freon 12-B2	75-61-6
1,1,2-Trichlorofluoromethane	Freon 11	75-69-4
Pentachloroethane	Ethane pentachloride	76-01-7
Chloropicrin	Trichloronitromethane	76-06-2
1,1,2-Trichlorotrifluoroethane	Freon 113	76-13-1
Hexachlorocyclopentadiene	HCCPD	77-47-4
1,2-Dichloropropane	Propylene dichloride	78-87-5
1,1,2-Trichloroethane	1,1,2-TCA	79-00-5
Trichloroethene	TCE	79-01-6
1,1,2,2-Tetrabromoethane	Acetylene tetrabromide	79-27-6
1,1,2,2-Tetrachloroethane	Acetylene tetrachloride	79-34-5
Hexachlorobutadiene	HCBD	87-68-3
o-Chlorotoluene	2-Chloro-1-methylbenzene	95-45-8
1,2-Dichlorobenzene	o-Dichlorobenzene	95-50-1
2-Chlorophenol	o-Chlorophenol	95-57-8
1,2-Dibromo-3-chlorpropane	DPCP	96-12-8
1,2,3-Trichloropropane	Allyl trichloride	96-18-4
Benzyl chloride	Chloromethylbenzene	100-44-7
p-Chlorotoluene	NA	106-43-4
Ethylene dibromide	1,2-Dibromoethane; EDB	106-93-4
1,2-Dichloroethane	Ethylene dichloride; 1,2-DCA	107-06-2
m-Chlorotoluene	NA	108-41-8
Bis(2-chloroisopropyl)ether	Bis(-chloroisopropyl)ether	108-60-1
1,2,4-Tetrachlorobenzene	1,2,4-TCB	120-82-1
Dibromochloromethane	Chlorodibromomethane	124-48-1
Tetrachloroethene	Perchlororethylene; PCE	127-18-4
trans-1,2-Dichloroethene	trans-1.2-DCE:	156-60-5
1,3-Dichlorobenzene	m-Dichlorobenzene	541-73-1
1-Chloro-1-nitropropane	Chloronitropropane	600-25-9
4-Chlorophenyl phenyl ether	p-Chlorodiphenyl ether	7005-72-3
cis-1,3-Dichloropropylene	cis-1,3-Dichloropropylene	10061-01-5
trans-1,3-Dichloropropene	trans-1,3-Dichlonopropylene	10061-02-0

6.5 EXPOSURE PATHWAYS

Exposure pathways for halogenated substances can occur through inhalation, ingestion, or skin absorption. Many of these substances can exert impact on human health through each route of entry individually or collectively. The exposure pathways for these substances are dependent on a variety of factors that include the following:

- The type of halogenated substance (e.g., halogenated hydrocarbon, halogenated solvent, halogenated organic).
- Deployment of engineering control and type.
- Personal protective equipment (PPE) usage.

Through the various exposure paths, the exposed individual can experience health conditions ranging from no health effect to severe health effect. Exposure avoidance remains the most effective way to protect consumers and workers from exposure and adverse effect from these substances.

6.5.1 Inhalation exposures and impact

Acute exposure to large doses of halogenated substances by inhalation can cause issues such as eye and respiratory tract irritation, contact dermatitis, central nervous system depression and unconsciousness, cardiac dysrhythmias, and hepatic and renal injury. Inhalation is the most common route of exposure for halogenated substances, from products such as paints, adhesives, cleaning solutions, and aerosolized insecticide sprays; from industries producing these solvents; and from contaminated waste disposal sites.[12] Although inhalation may be the primary route of exposure to halogenated substances, all the other exposure routes can play a role when it comes to exposure of these substances on people. A thorough exposure assessment will provide information on exposures that may occur during a work evolution and provide information of methods that can be taken to mitigate inhalation exposures.

6.5.2 Dermal exposures and impact

Halogenated substances can exert significant damage to the skin upon exposure. Two major skin disorders caused by halogenated substances are contact (photocontact) dermatitis and acneiform eruption. Photocontact dermatitis was traditionally caused by various halogenated salicylanilides, and chloracne (acne) is caused by halogenated aromatic hydrocarbons.[8] Work should be performed in a manner that skin exposure is avoided to prevent skin absorption and minor to severe damage. Tables 6.5 and 6.6

Table 6.5 Photocontact dermatitis – halogenated substances

Tetrachlorosalicylanilide (TCSA)	Dibromosalicylanilide (DBS, dibromosalan)
Tribromosalicylanilide (TBS)	Bithionol (thiobisdichlorophenol)
Trichlorocarbanilide (TCC, triclocarban)	Trifluoromethyldichlorocarbanilide (TFC)
Hexachlorophene	Chloro-2-phenylphenol (Dowicide 32)
Fenticlor (thiobischlorophenol)	Multifingin (bromochlorosalicylanilide, BCSA)
Jadit (buclosamide, butylchlorosalicylamide)	Triclosan
Chlorhexidine	Dichlorophene

Table 6.6 Chloracne–halogenated substances

Polyhalogenated naphthalenes	Polychloronaphthalenes
Polychromonaphthalenes	Polyhalogenated biphenyls
Polychlorinated biphenyls	Polybrominated biphenyls (PBBs)
Polyhalogenated dibenzofurans	Polychlorodibenzofurans, especially tri-, tetra-, penta-, and hemachlorodibenzofurans
Polybromodibenzofuran, especially tetrabromodibenzofuran	Contaminants of polychlorophenol compounds, especially herbicides
(2,4,5-T and pentachlorophenol) and herbicide intermediates	(2,4,5-troichlorophenol)
2,3,7,8-Tetrachlorodibenzo-*p*-dioxin	Tetrachlorodibenzofuran
Contaminants of 3,4-dichloroaniline and related herbicides (Propanil, Methazol, etc.)	3,4,3',4'-tetrachloroazoxybenzene
e,r,e',4'-tetrachloroazobenzene	

list some halogenated chemicals that are associated with photocontact dermatitis and chloracne.

Halogenated chemical compounds are important dermatitis causative agents in the field of occupational and environmental medicine. These two conditions also provide good experimental models for investigating dermatology[7] When working with halogenated substances, the user must operate with the assumption that if the product comes in contact with unprotected skin dermatitis may result.

6.5.3 Ingestion exposure and impact

Acute exposure to large doses of halogenated substances through ingestion can cause central nervous system depression and unconsciousness, cardiac dysrhythmias, and hepatic and renal injury.[14] Ingestion can be prevented by wearing the appropriate gloves when handling these substances, keeping hands clean, and avoiding eating in areas where these products are handled or stored.

6.6 CHEMICAL TO HUMAN INTERFACE

Human contact with halogenated substances can occur through many avenues including occupational, the air we breathe while outdoors, to our drinking water supply. Drinking water may contribute to exposure to halogenated solvents because of contaminated underground drinking water supplies. In general, these solvents are well absorbed by inhalation, dermal, or oral exposure. After absorption, small amounts may be exhaled in expired air, and the remaining amount can be rapidly distributed to tissues. Fatty tissues can accumulate these halogenated solvents that are slowly released back into the bloodstream. Eye and respiratory tract irritation may occur with exposure to high vapor concentrations of most halogenated substances. Allergic contact dermatitis has been reported following dermal exposure to 1, 2-dichloroethane. Exposures to vapor concentrations exceeding occupational exposure levels have resulted in fatigue, headache, delayed reactions, and neuropsychological impairment. Epidemiologic studies of workers exposed to various halogenated solvents have found occasional associations between exposure and reduction in fertility and spontaneous abortion in women.[13,14]

Many of the symptoms displayed by patients who have been exposed to halogenated substances resemble those that are seen in patients with other health effects. For example, workers exposed to exposure to 1-bromopropane (C_3H_7Br) can experience symptoms such as joint pain or leg weakness and pain leading to difficulty standing and walking; muscle twitching or numbness, tingling and prickling in the hands or feet, loss of vibration sense, anxiety, apathy, insomnia, and difficulties with concentration and memory.

6.7 EXPOSURE CONTROL METHOD

Implementing control methods to avoid exposure to halogenated substances can take on a variety of measures depending on usage and environment. For consumers of these products who use it at home, it is necessary that they understand the hazards of the products by consulting manufacturer's information and the safety data sheets (SDSs) for the product. To protect oneself and family, consider the following:

- Avoid using these products in closed space. Ensure that use occurs in a well-ventilated area. Use of product outside should be the preferred environment when feasible.
- Avoid skin contact. Protect skin by using gloves and long-sleeve clothing.
- Protect children by storing products in an area where it is not accessible to them.

- Purchase only what you need to use for the task.
- Read product label and adhere to hazard notification and precautions.

To prevent workers from overexposures, there are a variety of actions employers can take. Controlling exposures to halogenated compounds is multifaceted and begins with an exposure or risk assessment process that is effective in determining the risk and hazards to workers before beginning the work. The risk strategy must account for the evaluation of each substance and its potential hazard. For example, consider the following scenario: Two painters received a work order to paint a small office located in the welding shop. The office is traditionally used by the supervisors and workers during lunch and breaks. The office has a fully equipped bathroom. The work area is walkdown by the supervisor, workers, and a health and safety professional. The work is planned, and the appropriate work documentation is developed and approved. After participating in the walkdown, a risk/hazard assessment was documented (Figure 6.1) to determine and mitigate potential exposures that may occur during the task. In addition, to use the information gained from the risk assessment, other information was used to gain a complete understanding of potential hazards (e.g., methylene chloride and lead SDSs, PPE guide) (Figure 6.2).

6.7.1 Product elimination and substation

Product elimination and substation offers the most effective means to prevent hazards to consumers and workers. There have been instances where some halogenated substances have been discontinued for usage because of hazard posed to users. Elimination and substitution of hazardous products for less or nonhazardous chemical product such as "green" products continue to gain traction as focus are being placed on products that presents a hazard to general consumer. Although this method is the best way to protect consumers and workers from exposure to chemicals, it is not always feasible based on the application and availability of alternative products that are effective for the specific application.

6.7.2 Engineering controls

When product substitution cannot be used, engineering controls provide the next best means to control exposures to halogenated hydrocarbons. When designing systems and processes that require the use or production of halogenated compounds, engineering controls should be designed into the process. Engineering control in the form of exhaust ventilation is effective in removing the fume emitted by these substances and removing it from the environment and the breathing zone of workers. Other engineering control methods that should be considered involve placing a barrier between the

Location: Building X, Area AA   Date: XX – XX- XXXX

Task/Process description: Stripping paint from walls

Similarly Exposed Group (s): Painters

Potential Hazard	OEL	Exposure Route	Exposure Frequency	Exposure Duration	Controls	PPE	Exposure rating	Air sampling required (Y/N)
Methylene Chloride	25 ppm	Inhalation Skin contact	Continuous	7 hours	Exhaust ventilation	Half face organic vapor cartridge respirator, safety glasses	1	Y
Lifting	NA	NA	Twice daily	2 lifts	Mechanical lift and two-man rule	gloves	NA	N
Fall	NA	NA	Once daily	2 hours	Ladder and scaffolding	Hard hats, safety glasses	NA	N

Exposure rating scheme:

1- Continuous
2- Intermittent (2 – 3 times per day)
3- Infrequently (1 – 2 times per year)
4- Occasionally (1 – 2 times every 5 years)

Comments: stripper cans are 50 pounds each and 10 cans will be used.

Recommendations: Training for painters (lifting, HAZCOM, scaffolding, ladder safety, respirator). Medial approval to wear respiratory protection. Use mechanical means to lift and transport cans (pallet cart)

Future monitoring and evaluation: Determine monitoring frequency for future projects in similar environments using similar controls and PPE upon receipt of the laboratory results for samples collected for this evolution

Note: attached a copy of the monitoring strategy

Investigator/Assessor:

Title: Professional A

Investigator/Assessor signature:_____

Figure 6.1 Exposure assessment worksheet (example).

worker and the hazards such as enclosing the process or equipment with a box-type enclosure in concert with local exhaust ventilation. Deploying engineering controls for consumers at home or in their individual shops may not be feasible. However, any work involving the use of halogenated hydrocarbons should be conducted in well-ventilated areas and away from heat sources.

Sample identifier: XXXAAA-01   Date: XX-XX-2022

Sample Type (area, personal, blank) Personal

Collected by: Professional A   Collection time: 0930 am

Person/process sampled (name): Employee A

Work Environment

Process/task or work environment description: Stripping paint from walls

PPE used: Half face organic vapor cartridge respirator

Engineering controls used: Local exhaust ventilation

Temperature: 75 degrees F   Wind speed: NA   Wind Direction: NA
Humidity: 55%

Equipment and Calibration Data

Sampling equipment Mfg & model	Flow rate 3 L/min	Sample Pump Calibration Data:
Air sampling pump model A	Start: 9:30 am End: 3:30 pm Average: 420 min Volume: 1260	

Analysis and Results

Chemical/substance: Methylene chloride   Results concentration & unit: 1 PPM
Exposure Limit (OEL) 25 ppm

Analytical laboratory: Certified Industrial Hygiene Laboratory
Analyzed by: Laboratory Analyst B

Figure 6.2 Exposure monitoring worksheet.

6.7.3 Administrative controls

Administrative controls that include procedures, training, and policies are viable controls that are required by many regulations to assist employees in making reasonable decisions that will protect themselves from workplace harm. The effectiveness of most administrative controls is dependent on the individual implementing the control. Administrative control use for halogenated substances is a viable practice when implemented through training employees on the hazards of the substance and developing procedures and policies for employees to follow when working with or handling with these chemicals. Other administrative controls such as stay time are not a viable option to consider for usage with these chemicals.

6.7.4 Protective equipment and effectiveness

PPE should the last line of defense for protecting against chemical exposures. Widespread use of PPE occurs every day across the globe. PPE can be effective when the appropriate PPE is selected and used based on manufacturer's recommendations and requirements. In deploying PPE for halogenated compounds and achieving the intended effectiveness for protecting workers from exposure, knowledge of the following will be helpful:

- The chemical of concern and its chemical and physical properties.
- Hazards presented by the chemical and route of entry.
- Type of PPE that can provide the sought protection and their physical properties.

6.8 HUMAN BIOLOGICAL IMPACT

It has been shown by many authors, theorists, and researchers that halogenated compounds can have a negative health impact on humans. However, the data associated with humans are not available for a significant percentage of these chemicals. In an attempt to close the gap, researchers have conducted various studies with animals to glean data that can potentially be interpolated and used to provide some insight into whether there would be potential health impacts on humans to a wide array of these substances. Determining the biological impact of halogenated compounds on humans may not be simple for various reasons. For example, examining the levels of halogenated solvents in blood can reflect recent exposure. However, finding a measurable amount of any halogenated solvents in blood does not necessarily mean that the level of the solvent can cause an adverse health effect.[1]

6.8.1 Potential impact to exposed individual

Halogenated substances are a class of chemical substances that can not only result in biological hazards but also contribute to physical hazards to individuals as well. The physical hazards associated with the use of some halogenated substances come from the flammability and the high reactivity of the low-molecular-weight chemicals in the group. Higher-molecular-weight substances tend to be nonflammable. The associated health hazards of halogenated compounds are completely dependent on the specific chemical, route of exposure, the airborne concentration, and duration of exposure. Some examples are as follows:

- Halomethanes are derivatives of methane where one or more hydrogen atoms have been replaced with at least one halogen atom (F, Cl, Br, or I). These compounds when released in large quantities can act as asphyxiants when reducing the oxygen concentration in the air.
- Prolonged exposure to many chlorinated hydrocarbons through inhalation can and often result in liver and kidney damage. Exposure to unprotected skin to the solvents can result in defatting of the skin resulting in contact dermatitis. Methylene chloride and vinyl chloride have been shown to be human carcinogens.
- Inadvertent formation of phosgene is another hazard of concern from chlorinated hydrocarbons. Phosgene is used as an intermediate in the production of isocyanates and is formed when trace amounts of chlorinated degreasing solvents used to clean metals are exposed to the ultraviolet radiation that is emitted during arc welding. Phosgene is also formed by the thermal decomposition of chlorinated hydrocarbons for example, during a fire. Inhalation of phosgene in sufficient concentration can result in pulmonary edema and death[1].

The biological hazards from these chemicals can result in a large array of illnesses and injuries. Some health impacts are well documented in humans, and others are documented using animal research and extrapolated to potential impact on humans when exposed. The bottom-line is that halogenated substances can present a health hazard to consumers and workers.

6.8.2 Potential impact on offspring

Studies have been conducted that show some halogenated substances have the capability to adversely impact the fetus. The category of halogenated substances is large, therefore, as one would expect not all substances have been extensively evaluated to determine the potential health impact on a fetus. Studies have been conducted using animals to determine the impacts on animal fetus of various halogenated substances.[11]

Many solvents containing chlorine (called chlorinated solvents) have been linked to some level of increased reproductive risk. Many solvents also can be passed into breast milk.[15]Not all solvents are hazardous to the fetus; however, some studies have linked certain solvents with birth defects, miscarriages, low birth weight, and premature birth. The studies linking solvents to these problems are limited, and results are different because workers may use more than one solvent at a time, making it difficult to determine which solvent is causing the problem.

Excess fluoride exposure can have deteriorating effects on the expecting mother and fetus and is associated with some adverse pregnancy outcomes.

For example, females having elevated urinary fluoride levels were found in a study conducted by Goyal et al.[11] to have a strong association with the pregnancy complications such as anemia, miscarriage, abortion, and still birth[9] Excessive maternal bromide exposure during pregnancy from drugs and occupation have been reported to have adverse effects on the fetus and newborn, including central nervous system depression at birth and possible teratogenicity.[10]

6.9 TREATMENT OF RESULTANT ILLNESS

Treatment of disease may be difficult to diagnose in some people because exposure to halogenated compounds may manifest itself in symptoms that resemble other disease or issues in the human body. For example, let's consider 1-bromopropane, short-term exposure to the chemical can result in irritation of the eyes, nose, throat, and respiratory tract. Long-term exposure to 1-bromopropane can negatively affect peripheral nerves and the central nervous system. Symptoms that have been reported by patients include joint pain; leg weakness and pain making it difficult to stand and walk; muscle twitching or numbness; tingling in the hands or feet; loss of vibration sense; anxiety; apathy; insomnia; and difficulties with concentration and memory. Other health effects that can also occur include dermatitis, nausea and vomiting, diarrhea, difficulty in swallowing, disruption of menstruation, urinary difficulties, anemia or low red blood cell count, liver damage, and lung disease. Because of the variation in symptoms, the attending physician must be aware of the work environment for patients along with potential chemical exposures. This variation in symptoms can lead to misdiagnosis and prolong time in being able to identify the actual exposure chemical and potential route. A comprehensive risk and hazard assessment is helpful to assist physicians in treating employment-related exposures and injuries.

Treatment for illness and injuries resulting from halogenated compounds range from the application of topical skin cream or ointment to treat a minor rash to serious injuries or illness such as cancer. Treatment of complicated diseases like cancer is dependent on many variables that will ultimately guide the attending physician in determining the best of medical care.

REFERENCES

1. North Carolina Department of Labor (n.d.). Halogenated Hydrocarbons. NCDOL, North Carolina Department of Labor. https://www.labor.nc.gov/halogenated-hydrocarbons.
2. Reference Manual 021002 clean.doc, Appendix A (Halogenated Solvents List) and Appendix B (List of Carcinogenic Materials) (portland.gov).

3. Kirk Kenneth, L. (1991). *Biochemistry of the elemental halogens and inorganic halides*. Plenum Press. New York.
4. NIOSH Pocket Guide to Chemical Hazards (2005-149) 3rd printing (cdc. gov).
5. Fuge, R. (1988). Sources of halogens in the environment, influences on human and animal health. *Environmental Geochemistry and Health*, 10, 51–61. https://doi-org.ezproxy.proxy.library.oregonstate.edu/10.1007/BF01758592.
6. Odabasi, M. (2008). Halogenated volatile organic compounds from the use of chlorine-bleach-containing household products, *Environmental Science & Technology*, 42(5), 1445–1451, https://doi.org/10.1021/es702355u.
7. Manahan Stanley, E. (1994). Environmental chemistry, 6th ed., Lewis Publishers, Ann Arbor, MI.
8. Yamamoto, O., Toura, Y. (2003–08). Photocontact dermatitis and chloracne: Two major occupational and environmental sin disease induced by different actions of halogenated chemicals, *Journal of Dermatological Science*, 32 (2), P85–94.
9. Goyal, L.D., Bakshi, D.K., Arora, J.K., Manchanda, A., Singh, P. (2020). Assessment of fluoride levels during pregnancy and its association with early adverse pregnancy outcomes. *Journal of Family Medicine and Primary Care*, 9(6),2693–2698. https://doi.org/10.4103/jfmpc.jfmpc-213-20.
10. Miller M.E., Cosgriff J.M., Roghmann K.J. (1987). Cord serum bromide concentration: Variation and lack of association with pregnancy outcome. *American Journal of Obstetrics and Gynecology*, 157(4 Pt 1), 826–830. https://doi.org/10.1016/s0002–9378(87)80064-9.
11. Narotsky M.G., Brownie C.F., Kavlock R.J. (1997). Critical period of carbon tetrachloride-induced pregnancy loss in Fischer-344 rats, with insights into the detection of resorption sites by ammonium sulfide staining. *Teratology*, 56(4), 252–61. https://doi.org/10.1002/(SICI)1096–9926(199710)56:4<252: AID-TERA4>3.0.CO;2-0.
12. New Jersey Department of Health (2016). Hazardous Substance Fact Sheet: Iodine. NJ Health, New Jersey Department of Health. https://nj.gov/health/eoh/rtkweb/documents/fs/1026.pdf.
13. U.S. Environmental Protection Agency (2018). Appendix III to Part 268 - List of Halogenated Organic Compounds Regulated Under Â§ 268.32. Environmental Protection Agency. https://www.govinfo.gov/content/pkg/CFR-2018-title40-vol29/xml/CFR-2018-title40-vol29-part268-appIII.xml.
14. U.S. Department of Health and Human Services (2017). Other Halogenated Solvents. National Biomonitoring Program, Centers for Disease Control and Prevention, U.S. Department of Health and Human Services. https://www.cdc.gov/biomonitoring/OtherHalogenatedSolvents_BiomonitoringSummary.html
15. National Institute of Occupational Safety and Health (2019). Solvents – Reproductive Health. National Institute of Occupational Safety and Health, U.S. Department of Health and Human Services. https://www.cdc.gov/niosh/topics/repro/solvents.html.

Chapter 7

Inorganic compounds

7.1 INTRODUCTION

Inorganic compounds and substances cover a large variety of materials that are useful in enhancing the quality of life for humans. The list of inorganic compounds is very extensive, and it is very difficult to keep track of the many materials and their wide range of usage. The handbook of inorganic substances (2013) lists more than 1,00,000 inorganic substances or compounds that are used worldwide.[3,4,5]

These compounds are used in the production of a wide array of products from those used by consumers in the home, medical community, and industrial production processes. Inorganic compounds are those compounds that are not considered or classified as organic. These compounds tend to be made from materials such as ceramic, stone, and metal (made from rocks and minerals). Although useful for a wide array of applications, these materials can also pose a hazard to the environment as well as human health and well-being. Inorganic compounds are recognized as compounds not having a carbon–hydrogen connection or bond. In other terms, these chemicals do not contain both carbon and hydrogen or a carbon–hydrogen bond.

Inorganic substances are a collection of diverse chemicals classified based on the classical definition of substances that do not have carbon-based structures (with some exceptions, as discussed below). Within this class, groups are defined by structural similarities. The major groups of inorganic substances are as follows:

- **Metals:** In their "free" state, with the exceptions of gold, silver, platinum, and copper, most other metals are found in the environment as compounds. Some metals are known to be toxic, and some are essential nutrients at lower doses for humans and other organisms.
- **Carbon-containing inorganics:** Those compounds that are carbon-based that are grouped with inorganic substances because of their behavior and physicochemical properties are closely aligned with inorganic substances than to organic compounds. An example, hydrogen cyanide, widely used as an insecticide and produced as a chemical

DOI: 10.1201/9781003220114-7

reaction byproduct. Cyanides can also be found in the air because of automobile exhaust. Other carbon-containing inorganic substances include carbonates, carbonic acid, and oxides of carbon including carbon dioxide and carbon monoxide.

- **Fibers:** These are elongated, thread-like strands of molecules of various lengths that are often interwoven and entwined. For example, asbestos, a fibrous mineral that occurs naturally in certain rock formations. Asbestos has been banned for use in the United States because of its toxic properties.
- **Nutrients:** These are elements and compounds that provide essential nitrogen, phosphorous, potassium, and micro-nutrients for plants by direct application to the soil or by way of irrigation. For example, nitrogen and phosphorus are elements that occur naturally in aquatic ecosystems and support the growth of algae and aquatic plants, whereas nutrient over-enrichment in aquatic systems because of excess nitrogen and phosphorous contamination has been identified as a major water quality issue.[13,14]

Inorganics are also a component of the particulate matter that is emitted into the atmosphere across the United States. Sulfates and nitrates comprise the major inorganic portion of ambient PM2.5 in most areas of the United States. Sulfates tend to be more predominate in the eastern region, and nitrates tend to be predominate in the western region. Most of the sulfate derives secondarily from the oxidation of sulfur dioxide produced by natural sources, such as sulfur-containing gases emitted from, for example, oceans, wetlands, and volcanic emissions, and fossil fuel combustion sources. PM2.5-associated nitrate results predominately from the oxidation of nitrogen dioxide resulting mostly from combustion processes where fossil fuels are used. Nitric load in the environment can also result from natural oxidation by lighting and fires.[8]

Inorganic materials in the form of nanoparticles are extensively used for diagnostic and drug delivery purposes. These materials have distinctive features that are advantageous in developing imaging agents and theranostics. However, transmission from laboratory to clinic has been slow because of concerns of toxicity around in vivo and in vitro application. It has been demonstrated that these substances can contribute to unexpected toxicities in humans. Inorganic nanoparticles can potentially cause damage to cells, genetic materials, tissues, organs, and proteins because of their compositions and high-energy state.[1] Table 7.1 lists some common inorganic elements and substances.

Inorganic substances are commonly used in products that are used in everyday life for a wide array of applications. The household chemical industry is considerably large. For example, the market for cleaning products is presumed to worth, more than 4 billion a year in the United Sates

Table 7.1 Some inorganic elements and substances

Substance	CAS ID #	Characteristics
Antimony	7440-36-0	A silvery-white metal that is found in the earth's crust
Arsenic	7440-38-2	Naturally occurring element in the earth's crust. Inorganic arsenic compounds are mainly used to preserve wood
Barium	7440-39-3	A silvery-white metal that exists in nature in ores
Beryllium	7440-41-7	A hard, grayish metal found naturally in mineral rocks, coal, soil, and volcanic dust
Boron	7440-42-8	A compound that occurs naturally and is often found combined with other substances to form compounds known as borates
Cadmium	7440-43-9	A natural element in the earth's crust
Chlorine	7782-50-5	A yellow-green gas at room temperature that is heavier than air and has a strong irritating smell
Chromium	7440-47-3	A naturally occurring element embedded in rocks, animals, plants, soil, and volcanic dust and gases
Cobalt	7440-48-4	A naturally occurring element found in rocks, soil, water, plants, and animals
Copper Cyanide	7440-50-8	Occurs naturally in rocks, soil, water, and air
Fluorides	7782-41-4	A naturally occurring, pale yellow-green gas with a sharp odor (hydrogen fluoride and fluorine)
Hydrogen chloride	7647-01-0	A colorless to slightly yellow, corrosive, nonflammable gas at room temperature and is heavier than air with a strong irritating odor
Hydrogen cyanide	74-90-8, 143-33-9, 000151-50-8, 000592-01-8, 000544-92-3, 506-61-6, 0460-19-5, 506-77-4	A colorless to pale-blue liquid (hydrocyanic acid). A colorless gas at higher temperatures
Hydrogen fluoride	7664-39-3	A colorless, fuming liquid or gas with a strong, pungent irritating odor
Hydrogen sulfide carbonyl sulfide	7783-06-4	Occurs naturally in crude petroleum, natural gas, volcanic gases, hot springs, and result from bacterial breakdown of organic matter
Iodine	7553-56-2	Naturally occurring element found in sea water and in certain rocks and sediments
Lead	7439-92-1	A naturally occurring bluish-gray metal found in the earth's crust

(Continued)

Table 7.1 (Continued) Some inorganic elements and substances

Substance	CAS ID #	Characteristics
Manganese	7439-96-5	Naturally occurring metal that is found in rocks. Pure manganese is silver-colored, and does not occur naturally
Mercury	7439-97-6	Combines with other elements, such as chlorine, sulfur, or oxygen, to form inorganic mercury compounds that are usually white powders or crystals
Mercury metallic	7439-97-6	A dense liquid that vaporizes easily at room temperature
Nickel	7440-02-0	An abundant natural element that is hard, silvery-white metal
Radon	14859-67-7, 10043-92-2	An odorless and tasteless naturally occurring radioactive gas
Selenium	7782-49-2	A naturally occurring mineral element that is found in nature in most rocks and soils
Selenium hexafluoride	NA	A corrosive, colorless gas at room temperature
Silica	7631-86-9	An odorless solid compound composed of silicon and oxygen atoms
Silver	7440-22-4	A naturally occurring silver colored element
Strontium	7440-24-6	A naturally occurring element prevalent in rocks, soil, dust, coal, and oil
Sulfur dioxide	7446-09-5	A colorless gas with a pungent odor
Synthetic vitreous fibers	NA	A group of fibrous, inorganic materials that contain aluminum or calcium silicates, and are made from rock or stone, clay, slag, or glass
Thallium	7440-28-0	A bluish-white metal that is found in the earth's crust in small amounts
Thorium	7440-29-1	A radioactive substance that is naturally occurring
Tin and compounds	7440-31-5	A soft, white, silvery metal that is found naturally in the earth's crust
Titanium tetrachloride	7550-45-0	A colorless to pale-yellow liquid with fumes having a strong odor
Tungsten	7440-33-7	A naturally occurring element
Uranium	7440-61-1	A naturally occurring and radioactive substance
Vanadium	7440-62-2	Occurs in nature as a white-to-gray metal that is often found as crystals
Zinc	7440-66-6	One of the most common elements in the earth's crust and is found in air, soil, and water, and all foods

alone. These substances are generally used by consumers without thought of how they were produced or whether they belong to the class of products known as inorganic or organics. Consumer based inorganic products can also produce harm to humans and the environment if not used in a safe manner. Thus, it is pertinent that when using these products read the attached label that is designed to assist in safe product usage. This act can prevent a visit to the emergency room or a conversation with the poison hotline, which occurs more frequently that one would think.

Exposure to various environmental agents for many people begins in the very early stages of life during embryonic development (organogenesis). As such, the short- and long-term health implications of these exposures continue to be an area of interest. Henderson et al. published a report on the association between mothers' use of household chemical products during pregnancy and persistent wheezing in children. They also reported on a small adverse impact of these products on the lung function in children.

The study by Henderson et al. adds to the limited body of knowledge and published in the literature on the potentially harmful effects of chemicals used in the home on the respiratory health of children. Although the evidence is far from conclusive and a greater understanding of how these agents may contribute to disease is required; however, there seems to be sufficient evidence to caution people about the use of household chemical products, especially during pregnancy and in the presence young children.

To protect consumers, Federal legislation has been enacted that gives consumers information to help reduce the risks associated with the use of household products. In 1976, an amendment to the Federal Food, Drug, and Cosmetic Act (FDCA), was enacted that provides for the testing of drugs before they are marketed and requires presentation of data to the Food and Drug Administration (FDA) to show evidence of the efficacy and toxic side effects involved with each new drug to be introduced to the marketplace. Based on research, the FDA has restricted or prohibited the use of many chemicals that were deemed as potentially harmful to consumers such as vinyl chloride in aerosol products, chlorofluorocarbon propellants, and mercury-based cosmetics. The Federal Caustic Poison Act (FCPA) and the Federal Hazardous Substances Labeling Act (FHSLA) were passed in 1976, requiring that household products have a clearly marked label. To protect consumers product labelling must state in large letters "POISON" where applicable and must list the exact ingredients, the name and address of the manufacturer, the recommended medical treatment if the product is ingested, and directions for proper use and storage. In addition, the FCPA requires that various products have child-proof packaging to protect children.[7] Table 7.2 lists some inorganics that are commonly used in household across the globe.

Table 7.2 Household usage of inorganics

Inorganic compound	Chemical formula	Some usage
Ammonia	NH_3	Detergent, glass cleaners, dyes, fragrances,
Antimony	Sb	Decorative objects and tableware; making solder used in soldered joints that exist in most of the electronic products.
Calcium hydroxide	$Ca((OH)_2)$	Plaster used in interior wall construction
Carbon	C	Pencil, diamond jewelry, and diamond glass cutters
Copper	Cu	Decorative objects and tableware
Lead	Pb	Decorative objects and tableware
Magnesium hydroxide	$Mg((OH)_2)$	Antacid for neutralizing stomach acid and as a laxative
Salt	NaCl	Food seasoning, previously used to preserve food
Silica	SiO_4	Cat litter, a desiccant to control humidity, and some plastics
Silicon dioxide	SiO_2	Crystal jewelry, building sands, candle holders, and
Sodium hydroxide	NaOH	Found in cleaning products and in unblocking drains
Sodium hypochlorite	NaClO	Laundry bleaching and cleaning compounds
Sulfuric acid	H_2SO_4	Lead acid batteries to power vehicles
Tin	Sn	Decorative objects and tableware; making solder used in soldered joints that exist in most of the electronic products
Water	H_2O	Drinking water, maintaining the human body

In the past the pharmaceutical industry has predominately used organic drugs and have moved into using inorganic drugs for various reasons. Two such reasons that may have credence are as follows:

- May be transferred in the body by oxidation and ligand substitution reactions
- Some activities of metal ions have stimulated the development of metal-based therapeutics.

Worldwide use and sales of inorganic has grown rapidly over the past decade. Inorganic drugs are necessary and impactful to modern medicines because these drugs are used to diagnose and treat a variety of diseases and illnesses. Some inorganic compounds that are used for medicinal purposes are shown in Table 7.3. Inorganics such as ^{60}CO, ^{131}I ^{192}Ir are playing a significant role in nuclear medicine to treat cancer.[1]

Table 7.3 Inorganic compounds medicinal usages

Element	Substance	Medicinal use
Ag (silver)	Silver sulphadiazine	Antibacterial
Al (aluminum)	Aluminum hydroxide	Antiacid
As (arsenic)	Salvarsan, melarsen, tryparsamide	Antimicrobial
AU (gold)	Gold (I) thiolates Auranofin Au (I) diphosphine complexes	Antitumor Antiarthritic Antiviral
Ba (barium)	Barium sulfate	X-ray contrast
Bi (bismuth)	Bismuth subsalicylate, colloidal bismuth citrate, ranitidine bismuth citrate	Antiacid, antiulcer
Br (bromide)	Sodium bromide	Sedative
Cr (chromium)	Chromium complexes	Antidiabetic
Cu (copper)	Copper histidine complex	Supplement for Menkes disease treatment
Co (cobalt)	Coenzyme B_{12}	Supplement
Fe (iron)	Sodium nitroprusside Fe (III) desferrioxamine chelates	Vasodilator Antimicrobial
Gd (gadolinium)	Gd metallotexaphyrins	MRI contrast agent, Photodynamic therapy (PDT), radiopharmaceuticals
Hg (mercury)	Mercurochrome	Antiseptic
I (iodine)	I_2 Sodium iodine 131	Antiseptic Diagnosis of thyroid
Li (lithium)	Lithium carbonate	Manic depression
Lu (lutetium)	Lutetium complexes	PDT
Mg (magnesium)	Magnesium oxide	Antacid, laxative
Mn (manganese)	Manganese-superoxide dismutase (Mn-SOD) complexes	Superoxide scavengers MRI contrast agent
Pt (platinum)	Cisplatin, carboplatin	Anticancer
Ru (ruthenium)	Ru (III) complexes	Anticancer
Sb (antimony)	Pentostam, N-methylglucamine antimonate	Antileishmanial
Si (silicon)	Kaloinite $Al_2(OH)_4Si_2O_5$	Antidiarrheal
Sn (tin)	Tin (IV) ethyl etiopurpurin	PDT
Tc (technetium)	^{99m}Tc (V) propyleneamine oxime	Diagnostic imaging
Ti (titanium)	Titanocene dichloride, bis(β-diketonato) Ti (IV)	Anticancer
V (vanadium)	Bis (maltolato) oxovanodium (IV) bis (glycinato) oxovanodium (IV) bis (methylpicolinato oxovanadium (IV)	Antidiabetic
W (tungsten)	Polyoxometallates	Anti-HIV activity

(Continued)

Table 7.3 (Continued) Inorganic compounds medicinal usages

Element	Substance	Medicinal use
Zn (zinc)	Zinc oxide	Skin ointment
	Zn (II) bicyclam complexes	Antiviral
	Zinc citrate	Supplement
Zr (zirconium)	Zr (II) glycinato	Antiperspirant
Se (selenium)	Ebselen:	Synthetic antioxidant
	(2-phenyl-1, 2-benzisoselenazol-3	Anti-inflammatory
	(2H0-one	Neuroprotective agent
	Phenylaminoalkyl selenide:	Antihypertensive
	(4-hydroxy-α-methyl-phenyl-	Antineoplastic and antiviral
	2aminoethyl selenide)	agent
	Selenazofurin	Antiallergic agent
	Seenotifen	

Many inorganic substances are necessary to maintain human health. The human body needs approximately 20 essential elements to function properly that belong to the class of organic and inorganic.[6] These substances that are critical to human health are referred to as minerals. Minerals are inorganic compounds found in food that work with other nutrients to ensure optimal function of the human body.[16,17] Some of the inorganic substances and their importance are listed in Table 7.4 as an example of the role and importance of inorganic substances to human health.

Toxic substance portal (TSP) highlights many inorganic substances along with hazards and health impacts. The TSP is a good resource to use to help assess the risk and hazards of these substances to workers. The labels found on household products are a good source of information for consumers to use in assessing safe product usage. Many of these substances are used in homes, medicinal purposes, and industries in the production or development of consumer products. In addition, some of these substances used in small amounts have health benefits that are necessary for everyday life. On the other hand, some of the same substances that are useful can also be harmful when exposed to a high level. Because of the wide span of using the potential of exposure is increased amount the populous.

7.2 EXPOSURE PATHWAYS

Inorganic substances can be naturally occurring or produced intentionally. When produced intentionally, these substances are generally incorporated into the production of products. When produced unintentionally, these products are generally released as byproducts. Inorganic is a highly general classification that refer to large and varied groups of substances, and so generalizations regarding the use and common routes of exposure are difficult to make for this broad class.[14]. Many inorganic substances can enter

Table 7.4 Some inorganic substances needed for healthy lifestyle

Mineral	Health purpose	Deficiency symptoms
Calcium	Maintain bone structure and health; nerve and muscle functions to include cardiac function	Slower growth rate, weak and brittle bones
Chloride	Balances body fluids, digestion	Loss of appetite, muscle cramps
Chromium	Insulin function	High blood sugar, triglyceride, and cholesterol balance
Cobalt	Component of B$_{12}$	None
Copper	Act as an antioxidant, red blood cell production, nerve and immune system function, collagen formation, acts as an antioxidant	Anemia, low body temperature, bone fractures, low white blood cell count, irregular heartbeat, and thyroid problems
Fluoride	Maintain bone and tooth structure	Increased cavities, weak bone structure and teeth
Iodine	Thyroid function	Hypothyroidism: fatigue, weight gain, dry skin, sensitivity to temperature
Iron	Transport of oxygen in blood, production of ATP	Anemia, weakness, and fatigue
Magnesium	Enzyme activation, production of energy	Agitation, anxiety, problems sleeping, nausea and vomiting, abnormal heart rhythms, low blood pressure, muscular problems
Manganese	Formation of connective tissue and bones, blood clotting, development of sex hormone, metabolism, brain and nerve function	Infertility, bone malformation, weakness, and seizures
Phosphorous	Bone formation, metabolism, ATP production	Rare
Potassium	Nerve and muscle function	Hypokalemia: weakness, fatigue, muscle cramping, gastrointestinal and cardiac problems
Selenium	Antioxidant, thyroid function, immune system function	Muscle pain
Sodium	Blood pressure, blood volume, muscle and nerve function	Rare
Sulfur	Component of amino acids	Protein deficiency
Zinc	Immunity, reproduction, growth, blood clotting, insulin, and thyroid function	Loss of appetite, poor growth, weight loss, skin issues, hair loss, vision problems, lack of taste or smell

the body and produce harmful effects through all routes of entry. Routes of exposure can include ingestion from drinking water containing, for example nutrients and metals; through the skin from coming in contact with, for example, contaminated soil with carbon-containing inorganics and metals, and through inhalation of carbon monoxide and fibers.[14]

The primary exposure pathways for inorganic materials may be different depending on the substance. However, there are exposure pathways, such as inhalation and dermal, that are germane to all inorganic substances. Among the more than 1,00,000 identified inorganic substances, many of these substances have the capability to cause harm to humans through all routes of entry. Some pathways may produce greater damage to the body depending on a variety of circumstances such as chemical type, length of exposure, the amount of chemical, and target organs.

7.2.1 Inhalation exposures and impact

Inhalation of fumes and fibers from organic substances should be considered a viable pathway for many inorganic substances especially those that are associated with workplace usage. The use of synthetic, inorganic fibers is extensively available and used in consumer and industrial applications including textiles, plastics, construction, and automotive materials. Examples of use in the industries include insulation materials, cements, steel fiber composites, sealants, rubbers, and electronics cables. Inhalation of fibers, for workers, is of concern because the size and shape of fibers has the capability to penetrate deep into the lungs after inhaled.[14] Exposure to inorganic particles may induce fibrosis in the lung (Table 7.5).[11] Exposure to fumes can cause a variety of health issues ranging from breathing pathway irritation to cancer. List some chemical irritants that can cause acute injury through inhalation.

7.2.2 Dermal exposure impact

Dermal exposure to inorganics in dust could occur by an individual coming in contact with dust that has settled on surfaces such as floors, clothing, or other surfaces. Therefore, dermal exposure should be of concern for all inorganics. This is true for household use as well as ones used in the work environment. In fact, if one were to read the label affixed to the household chemicals used, one will discover a caution to avoid skin and eye contact for many of these substances. Dermal exposure is a significant concern with exposure to unprotected skin, which is a possibility. When working with or using inorganics, care must be given to avoid shin contact as many of these substances can cause significant impact on the skin and in some cases depending on the level of exposure in internal organs. To illustrate this point, let's consider ammonium thiocyanate. The new Jersey Department of Health and Senior Services publishes a Hazardous Substance Fact alerting

Table 7.5 Some inorganics substances and potential impacts

Substance	Potential health impact
Antimony	Has been used as a medicine to treat people infected with certain types of parasites. Some side effects have been reported, including heart problems, nausea and vomiting, and muscle and joint pain. Inhaling antimony dust can cause heart and lung problems, stomach pain, diarrhea, vomiting, and stomach ulcers. Antimony can also irritate the skin and eyes.
Arsenic	Inhaling high levels of inorganic arsenic can result in a sore throat or irritation of the lungs. Ingesting extremely high levels of arsenic can result in death. Exposure to lower levels can cause nausea and vomiting, decreased production of red and white blood cells, abnormal heart rhythm, damage to blood vessels, and a sensation of "pins and needles" in hands and feet. Ingesting or breathing low levels of inorganic arsenic for a long time can cause a darkening of the skin and the appearance of small corns or warts on the palms of the hands, soles of the feet, and torso. Skin contact may cause redness and swelling.
Barium	The health effects of the different barium compounds is dependent on how well the compound dissolves in water or in the contents of the stomach when ingested. Can potentially cause, vomiting, abdominal cramping, diarrhea, difficulties in breathing, changes in blood pressure, gastrointestinal disturbances, and muscular weakness when exposures exceed EPA drinking water standards for short periods of time. Consuming large amounts can lead to heart rhythm, paralysis, and potentially death.
Beryllium	Can be harmful if inhaled. An acute condition called acute beryllium disease, beryllium sensitization, and chronic beryllium disease can result if beryllium air levels are high enough. Some symptoms include feeling of weakness, difficulty in breathing, anorexia, weight loss, heart enlargement, and heart disease. Beryllium contact with skin that has been scraped or cut may cause rashes or ulcers.
Boron	People working in environments where a high level of dust exists, and borates are mined and processed can experience irritation of the nose, throat, and eyes. Exposure to large amounts of boron over short periods of time can affect the stomach, intestines, liver, kidney, and brain and can eventually lead to death.
Cadmium	Inhaling high levels of cadmium can severely damage the lungs. Eating food or drinking water with high levels can irritate the stomach, leading to vomiting and diarrhea. Long-term exposure to low levels of cadmium in air, food, or water can lead to a buildup of cadmium in the kidneys, lung damage, and fragile bones.
Chlorine	Exposure to low levels of chlorine can result in nose, throat, and eye irritation; at higher levels, inhaling chlorine gas can result in changes in breathing rate, coughing, and damage to the lungs. Hypochlorite solution coming in contact with the skin can produce irritation.
Chromium	Chromium(III) is an essential nutrient that helps the body use sugar, protein, and fat. Breathing high levels of chromium(VI) can cause irritation to the lining of the nose, nose ulcers, runny nose, and breathing problems including asthma, cough, shortness of breath, or wheezing. Skin contact with chromium(VI) compounds can cause skin ulcers.

(Continued)

Table 7.5 (Continued) Some inorganics substances and potential impacts

Cobalt	Beneficial for humans because it is part of vitamin B12. Exposure to high levels of can result in lung and heart effects and dermatitis. Exposure to large amounts of radiation from radioactive cobalt can cause cell damage from radiation. Although rare, radiation syndrome may be experiences that may include nausea, vomiting, diarrhea, bleeding, coma, and even death
Copper	Essential for good health. Inhaling high levels of copper can cause irritation of the nose and throat. Ingesting high levels of copper can cause nausea, vomiting, and diarrhea. At extremely high levels liver and kidney damage can occur and even death.
Cyanide	Exposure to high levels of cyanide for a short time harms the brain and heart, can even cause coma and death. Inhalation of low levels of hydrogen cyanide over a period of years can result in breathing difficulties, chest pain, vomiting, blood changes, headaches, and enlargement of the thyroid gland.
Fluorides	Small amounts of fluoride help prevent tooth cavities; however, high levels can negatively impact health. In adults, exposure to high levels of fluoride can have an adverse effect on the bone. Fluorine and hydrogen fluoride irritate the skin, eyes, and respiratory tract. At high levels, hydrogen fluoride may also damage the heart.
Hydrogen chloride	Irritating and corrosive to any body tissue. Brief exposure to low levels can cause throat irritation. Exposure to higher levels can result in rapid breathing, narrowing of the bronchioles, blue coloring of the skin, accumulation of fluid in the lungs, and even death. Long-term exposure to low levels can cause respiratory problems, eye and skin irritation, and discoloration of the teeth.
Hydrogen cyanide	Hydrogen cyanide is absorbed through inhalation and can produce death within a matter of minutes. Absorption can occur through skin contact if vapor concentration is high or with direct contact with solutions.
Hydrogen fluoride	Hydrogen fluoride exposures mostly occur through inhalation of the gas and dermal contact with hydrofluoric acid. Hydrofluoric acid is a serious systemic poison. It is highly corrosive.
Hydrogen sulfide Carbonyl sulfide	Human studies suggest that the respiratory tract and nervous system are the most sensitive targets of hydrogen sulfide toxicity. Exposure to low concentrations of hydrogen sulfide can cause irritation to the eyes, nose, or throat; can cause headaches, poor memory, tiredness, and balance problems. Little is known about the health effects of exposure to carbonyl sulfide.
Iodine	Iodine is needed by your thyroid gland to produce thyroid hormones. Exposure to high levels of nonradioactive and radioactive iodine can damage the thyroid impacting other parts of your body such as skin, lung, and reproductive organs
Lead	Lead can affect almost every organ and system in the body. The nervous system is the main target for lead poisoning in children and adults. Long-term exposure can result in decreased learning, memory, and attention, and weakness in fingers, wrists, or ankles. Lead exposure can also cause low iron in the blood and damage to the kidneys and increases in blood pressure. Exposure to high lead levels can damage the brain and kidneys and can cause death

(Continued)

Table 7.5 (Continued) Some inorganics substances and potential impacts

Manganese	Exposure to high levels of manganese include behavioral changes and other nervous system effects, lung irritation and reproductive effects
Mercury	Exposure to high levels of metallic, inorganic, mercury can damage the brain, kidneys, and developing fetus. Effects on brain functioning may result in irritability, shyness, tremors, changes in vision or hearing, and memory problems. Short-term exposure to high levels of metallic mercury vapors can cause lung damage, nausea, vomiting, diarrhea, increases in blood pressure, skin rashes, and eye irritation
Mercury metallic	Exposure to lower levels of airborne mercury for prolonged periods of time may produce effects, such as irritability, sleep disturbances, excessive shyness, tremors, coordination problems, changes in vision or hearing, and memory problems. Exposure to high levels of metallic mercury vapor can cause, coughing, chest pains, nausea, vomiting, diarrhea, increased blood pressure, skin rashes, eye irritation, brain, kidney, and lung damage and may seriously harm a developing fetus
Nickel	The most common harmful health effect of nickel in humans is an allergic reaction manifesting itself in a skin rash at the point of contact. Inhaling high levels can produce chronic bronchitis and reduced lung functions. Ingesting high levels may result in stomach cramps, blood, and kidney effects.
Radon	Scientists believe that the alpha radiation dose from long-term exposure to elevated levels of radon in air can increase the chance of getting lung cancer.
Selenium	Short-term oral exposure to high concentrations of selenium can cause nausea, vomiting, and diarrhea. Chronic oral exposure to high concentrations of selenium compounds can produce a disease called selenosis with symptoms of hair loss, nail brittleness, and neurological abnormalities. Brief exposures to high levels of elemental selenium or selenium dioxide in air can result in respiratory tract irritation, bronchitis, difficulty breathing, and stomach pains. Longer-term exposures cause respiratory irritation, bronchial spasms, and coughing.
Selenium Hexafluoride	Exposure to low levels of selenium hexafluoride gas can cause irritation of the respiratory tract, skin, and eye. Exposure to high levels can lead to severe skin and eye damage, accumulation of fluid in the lungs, and eventually death. Long-term exposure to low concentrations of selenium hexafluoride may cause effects such as pale appearance, nervousness, depression, gastrointestinal problems, lack of appetite, weight loss, and dental defects. Exposure to the liquefied gas can cause severe skin injury, frostbite, severe eye damage and potentially blindness
Selenium hexafluoride	Exposure to low levels of selenium hexafluoride gas can cause irritation of the respiratory tract, skin, and eye. Exposure to high levels can lead to severe skin and eye damage, accumulation of fluid in the lungs, and eventually death. Long-term exposure to low concentrations of selenium hexafluoride may cause effects such as pale appearance, nervousness, depression, gastrointestinal problems, lack of appetite, weight loss, and dental defects. Exposure to the liquefied gas can cause severe skin injury, frostbite, severe eye damage and potentially blindness
Silica	Can cause respiratory diseases

(Continued)

Table 7.5 (*Continued*) Some inorganics substances and potential impacts

Silver	Exposure to high levels of silver for a long period of time may result in a condition known as argyria, blue-gray discoloration of the skin and other body tissues. Exposure to high levels of in the air can resulted in breathing problems, lung and throat irritation, and stomach pains. Skin contact can cause skin rash, swelling, and inflammation.
Strontium	Breathing high levels of radioactive strontium can damage bone marrow, cause anemia, and prevent the blood from clotting properly.
Sulfur dioxide	Exposure to high levels can be life-threatening. Inhaling high levels can result in burning of the nose and throat, breathing difficulties, and severe airway obstructions.
Synthetic vitreous fibers	When synthetic vitreous fibers are airborne, they can cause irritation of the eyes, nose and throat, and the lung. When these fibers contact the skin, they can also cause irritation.
Thallium	Inhaling high levels of thallium can result in numbness of fingers and toes. Ingesting high can result in vomiting, diarrhea, temporary hair loss, and effects on the nervous system, lungs, heart, liver, kidneys, and potentially death.
Thorium	Breathing high levels of thorium dust can result in an increased potential of getting lung disease.
Tin and compounds	Ingestion of large amounts of inorganic tin compounds can cause stomach pains, anemia, and liver and kidney problems. Inhaling or ingesting, or skin contact with some organotins, such as trimethyltin and triethyltin compounds, can interfere with the way the brain and nervous system work and potentially death in severe cases. Compounds coming in contact with the skin or eyes can cause irritation.
Titanium tetrachloride	Titanium tetrachloride can produce irritation to the skin, eyes, mucous membranes, and lungs. Exposure to liquid titanium can result in skin burns.
Tungsten	Contact with tungsten can irritate the skin and eyes. Repeated exposure to very high levels of tungsten may affect the lungs and mucus membranes.
Uranium	When inhaled or ingested natural uranium and depleted uranium can cause kidney damage.
Vanadium	Exposure to high levels of vanadium pentoxide in air can result in nausea, diarrhea, stomach cramps, and lung damage.
Zinc	Zinc is an essential element in our diet. Too little zinc can cause problems and too much zinc is also harmful. Large doses taken by mouth even can cause stomach cramps, nausea, vomiting, and anemia and decrease the level of good cholesterol. Inhaling large amounts of zinc (as dusts or fumes) can cause metal fume fever. Skin irritation is probable in humans.

the public of the hazards surrounding the use of this substance. Ammonium thiocyanate has many usages in chemical manufacturing and photography. It can irritate the skin and eyes upon contact.[17] As with ammonium thiocyanate, most inorganics can produce similar effect to the skin upon exposure. One way to prevent skin exposure is to ensure that the skin is covered with the appropriate type of material that is resistant to the particular inorganic being used. Personal protective equipment (PPE) being the last resort for exposure control can be effective in preventing skin contact and therefore avoid negative health impacts. There are a variety of gloves available to serve as a barrier between the hands and arms and the chemical of concern.

7.2.3 Ingestion exposure and impact

Consumer usage of inorganics potentially represents the largest opportunity for exposure through the ingestion route. Thus, it is important for users to understand the hazards of these products, the risk associated with usage, and measure that is required to prevent consumption. Ingestion potentially represents the most important pathway for children exposure to the diverse amounts of these chemicals that are used in homes across the globe. The National Poison Control Center reported for 2019 that 10.5% of the poison cases reported for that year resulted from cleaning substances. Good hygiene practices play an important role in preventing the consumption of these materials. Ingestion of inorganics is preventable; therefore, when these products are used, plans should be put into place to prevent hazards to consumers and workers.

7.3 CHEMICAL TO HUMAN INTERFACE

Inorganic substances can come in contact with humans during their daily life living as these substances occur in multiple environmental media, including air, water, and soil. Some of these substances can also accumulate in living organisms such as fish or other types of food. To demonstrate the actions and transport of inorganics in the different media using two medals that have been studied extensively, lean and mercury, related sources are detailed below.

Air

- Burning coal produces emissions of elemental mercury and divalent mercury. Divalent mercury that deposits in surface water can be transformed to methylmercury by anaerobic microbes.
- Lead deposited in the air has the capability of traveling long distances before depositing onto soil or in water sources.

Food

Lead and mercury substances can bioconcentrate in plants and animals used for human food sources. Soil

- Lead can adsorb and affix soil particles and unlikely migrate to groundwater from soil.

Water

- Lead can affix to sediment particles in surface water, where it can remain for many years.
- Lead contamination of drinking water could occur after treatment, for example, from corrosion of lead plumbing materials.[18]

7.4 EXPOSURE CONTROL METHODS

Exposure to inorganics in the workplace remains the primary responsibility of the employer. However, when it comes to the inorganics, we used in our homes it is the responsibility of the individual to protect themselves and their families from exposures. This section will focus on the protection of workers while in the workplace. The first step in controlling exposures is to evaluate the potential hazards through the completion of a risk or hazard assessment. The forms shown in Figures 7.1 and 7.2 can be helpful in conducting the hazard assessment.

Additional information on conducting a risk assessment can be found in Appendix A. Information on conducting a monitoring campaign is found in Appendix B.

7.4.1 Product elimination and substation

Product elimination and substitution is always the best approach to controlling exposures to various substances. In the case of inorganics, there are some opportunities to use safer or greener products for some household application, consumers accepting the concept of product substitution may be hampered by their belief that only the pure product may be deemed effective for use and all other products that may be safer to use is not viewed as being as effective. When used in the industrial work environment, substitution is not always feasible. However, considerations should be given to removing hazardous substances from the work environment where feasible. Other control methods become necessary to prevent exposures to inorganics when product substitution is not feasible.

Location: _____   Date: _____

Task/Process description: _____
_____
_____

Similarly Exposed Group (s): _____

Potential Hazard	OEL	Exposure Route	Exposure Frequency	Exposure Duration	Controls	PPE	Exposure rating	Air sampling required (Y/N)

Exposure rating scheme:

1- Continuous
2- Intermittent (2 – 3 times per day)
3- Infrequently (1 – 2 times per year)
4- Occasionally (1 – 2 times every 5 years)

Comments:

Recommendations:

Future monitoring and evaluation:-
_____
_____

Note: attached a copy of the monitoring strategy

Investigator/Assessor:

Title: _____

Investigator/Assessor signature: _____

Figure 7.1 Exposure assessment worksheet (example).

7.4.2 Engineering controls

The best option for protecting workers against inorganics is to design safety into the process and systems. Engineering control when effective will eliminate or minimize worker exposure to chemicals, thereby minimizing exposure potential. However, when considering usage of inorganics in the home environment, engineering controls are generally not a consideration. The average consumer when using products designed for home usage may

Sample identifier: _____   Date: _____

Sample Type (area, personal, blank) _____

Collected by: _____   Collection time: _____

Person/process sampled (name):

Work Environment

Process/task or work environment description: _____
_____

PPE used: _____

Engineering controls used: _____

Temperature: _____ Wind speed: _____ Wind Direction: _____
Humidity: _____

Equipment and Calibration Data

Sampling equipment Mfg & model	Flow rate Start: _____ End: _____ Average: _____	Units sample start time:	

Analysis and Results

Chemical/substance	Concentration & unit:	Exposure Limit (OEL)
Analytical laboratory: _____ Analyzed by: _____		

Figure 7.2 Exposure monitoring worksheet.

assume that the produce is safe if used as prescribed. Engineering controls to assist with controlling exposures to inorganics include laboratory hoods, ventilated enclosures, local exhaust ventilation system, and tight-fitting lids or fixture to prevent vapor from escaping. When using these control methods, a maintenance program must be developed and implemented to ensure that these measures are and remain effective for its intended purpose and usage as a safety control measure.

7.4.3 Administrative controls

Administrative controls continue to be contemplated and implement by many employers to control exposures to a wide array of substances. Although not the most preferred method, can be used successfully when used with other controls and designed strategically with the appropriate level of knowledge. Inorganic compounds are widely used, and this wide array of use adds complication to the ability to control exposure for a large population. It is easier to control exposures administratively when used in the work environment. Consumers using these chemicals in home is not likely to automatically think of the potential hazards and seek to implement controls beyond the ones recommended on the label. Therefore, from an administrative control perspective, it is important that manufactures include as much information on ways consumers can protect themselves.

7.4.4 Protective equipment and effectiveness

PPE are used frequently to control exposures to inorganics. PPE is used by consumers in the home environment as well as by employers when protecting workers from inorganic exposures. Before using PPE, it is necessary to have knowledge of the hazards in which protection is being obtained. In the home environment, it is reasonable that gloves will be used to protect the hands. When selecting a glove to protect against skin exposure, consult a glove chart or manufacture to ensure selection of the appropriate glove. In the work environment because the diversity of hazards that may be involved in any one task or process involved in the use of inorganics, it is recommended that an evaluation be conducted to ensure that all PPE needs are identified and the appropriate PPE is selected. Appendix B offers more information on the use and selection of PPE. An example of a completed PPE evaluation form is shown in Figure 7.3.

7.5 HUMAN BIOLOGICAL IMPACT

The impact on humans with regard to exposure to inorganics depend on the chemicals and their target organ. These substances have the capability to alter the function of the human body and create condition where disease can develop and even death. Exposure to some inorganics can severely impact the human body in various ways. Some of the biological impacts are listed in Table 7.6. The objective is to know the hazards associated with inorganics and avoid exposure to high levels for prolonged period of time. In addition, the biological impact of many inorganics on humans have not been studied completely, and in many cases, studies have been done in other

Performed by: _____   Date:_____

Process/Task: _____   Work group:_____

Source	Hazard	PPE Required	Special instructions
Chemicals ☐ Biological Agent ☐ Physical process ☐	Eye or Face injury potential Impact from flying objects ☐ Chemical splash in eyes ☐ Facial skin chemical contact ☐ Skin contact potential Hand ☐ Body - whole ☐ Extremities ☐ Face ☐	Safety glasses ☐ Safety glasses with side shields ☐ Safety goggles ☐ Face shield ☐ Latex gloves ☐ Rubber gloves ☐ Chemical resistant gloves ☐ Chemical apron ☐ Tyvek ☐ Lab coat ☐	
Task generating airborne fibers, dust, fumes, mist, or vapor ☐	Inhalation exposure above OEL ☐ Inhalation exposure below OEL ☐	Respiratory protection Organic vapor cartridge ☐ Particulate cartridge ☐ Other ☐	Request exposure assessment to be completed of the airborne constituents
High noise from equipment or task performance ☐	Inhalation exposure above OEL ☐ Inhalation exposure below OEL ☐	Ear plugs ☐ Earmuffs ☐ Other ☐	Noise survey to verify level and PPE requirement
Non ionizing radiation Lasers ☐ Welding ☐	Potential burns: Eyes ☐ Skin ☐	Laser safety glasses ☐ Welder's helmet ☐ Other ☐	Request a non-ionizing survey
General safety – physical hazards resulting from equipment, process, or material ☐	Foot injury potential ☐ Impact potential – head ☐ Electrical ☐	Safety shoes ☐ Hard hat ☐ Cut resistant gloves ☐ Electrical gloves ☐ Other ☐	
Other ☐	Cold or heat extremes ☐	Thermal gloves/clothing ☐	

Figure 7.3 PPE selection form.

species that may not be completely transferrable to human biological health impact.

Consider inorganic mercury as an example and its biological impact on the human body. Inorganic mercury compounds can be found in two oxidative states known as mercurous and mercuric. These substances are generally in solid states as mercurous or mercuric salts and mercury compounds with chlorine, sulfur, or oxygen. Most exposures to humans are to mercury in a chronic and low-dose fashion. However, exposure to high levels of mercury can occur in industrial accidents for very short periods. Once exposed, approximately 7% to 15% of doses of inorganic mercury compounds are absorbed in the gastrointestinal tract after ingested. Inorganic mercury can also be exposed to the skin through the transport of mercury across the epidermis and by the sweat glands, sebaceous glands, and skin hair follicles. The highest concentration of inorganic mercury is generally in the kidney, which is a major target organ for the compound.

Acute high-dose exposure of mercuric salts because of its corrosivity effect causes burning in the chest pain, darkened discoloration of the mucous membrane, and severe gastrointestinal symptoms. Mercury salts are irritants on the skin and cause dermatitis, discoloration of the nails, and corrosion of the mucous membranes. Chronic inorganic mercury poisoning is rare and occurs with pure inorganic mercury salts.

Inorganic mercury creams and ointments have been used as antiseptics for decades. Cosmetic soaps and creams containing inorganic mercury

Table 7.6 Some inhalation hazards

Substance	Source of potential exposure
Ammonia	Agriculture (mostly fertilizers); plastics, pesticides, explosives, and detergents; refrigerants, and home cleaning products
Chlorine	Household cleaners (household accidents involving the inappropriate mixing of hypochlorite cleaning solutions with acidic agents), paper production, sewage treatment, swimming pool maintenance, chemical manufacture, disinfection, chemical warfare
Hydrogen chloride	Dyes, fertilizers, textiles, rubber manufacture; metal ore refining; meat wrappers
Hydrogen fluoride	Phosphate fertilizer, metal refining and etching, glass and ceramic etching, microelectronic, masonry, pharmaceuticals, chemical manufacture; rust removal agents
Hydrogen sulfide	Decaying organic matter, in sewer and barns; petroleum refining, viscose rayon, rubber and mining industries; hot-asphalt paving
Sulfur dioxide	Airway pollution, burning of oil and coal, smelting, power plants, wineries, paper manufacture, chemical manufacture, and food preparation

compounds have been produced in several countries. Women in some countries use these creams as skin-lightening products and cosmetics for skin lightening and freckle removal or prevention. Mercury poisoning after using skin-lightening products has been reported by several countries such as Africa, Europe, the United States, Mexico, Australia, and China. Mercury salts can inhibit melanin formation resulting in skin lightening. Strict prohibition on mercury use in cosmetic products, comprehensive assessment of mercury contamination in products imported from other countries, and educating the public on the adverse health effects of products containing mercury is necessary to curve or prevent exposures.

Blood mercury level provides useful exposure information when measured soon after a short-term and high-level exposure evolution; however, the level decreases rapidly within days after exposure has occurred. Estimation of mercury concentration in the urine is the best biomarker to use to quantify long-term exposure and provides an indicator of body burden to inorganic mercury. Mercury found in the urine comes directly from the mercury deposited within the tissues of the kidney. Mercury analysis of hair can also be useful for determining chronic exposure because of the abundant sulfhydryl groups in hair; however, it is not recommended for biological monitoring for internal impacts.

The priority in the management of mercury poisoning is to stop exposure to mercury immediately by removing the person or eliminating the source. Mercury excretion can be increased by using chelating agents, such as dimercaprol, penicillamine, and 2,3-dimercapto-1-propanesulfonic acid. Although mercury poisoning is somewhat rare today, the risk to the health of humans continues to be of public concern because of the wide distribution and continuous contamination of mercury in the environment from natural and anthropogenic sources.[9]

7.5.1 Potential impact on exposed individual

Exposure to inorganic contaminants can occur through various avenues such as ingestion of contaminated food or water, ingestion of soil, inhalation of soil particulates, and dermal contact. Some members of the populations may experience increased exposures to the inorganic contaminants due to behavior, or occupation. These populations are inclusive of breast-feeding infants, fetuses, children, and people working in various occupations. Some examples are as follows:

- Infants that are breast feeding may potentially be exposed to methyl-mercury if the mother has been exposed.
- Young children may be exposed to lead dust if they are in an environment where lead dust is present, and their hands become contaminated with the dust and the toddlers in turn place their hands in the mouth.
- Workers may be exposed to inorganic contaminants depending on their occupation.[15]

When exposed to inorganics, the impact on the exposed individual is substance-dependent on a cadre of factors including the following:

- Route of entry
- Amount of the substances
- Concentration of the substance
- Current health of the exposed individual
- Length of time exposed
- Environmental conditions
- Other factors

Because the list of inorganics is extensive with the products having the capability to produce health effects from contact dermatitis to cancer, it is advisable that users of these products

- Take precautions and assume that the worse can happen when using.
- Educate themselves on the health effects of the product.
- Understand the need and type of PPE that may assist in exposure prevention.
- Know sign and symptoms of exposure.
- Must have a program in place to protect workers from potential exposures.

A list of some inorganic substances along with their health impact on exposed individual is listed in Table 7.6. The table is not comprehensive and represents only examples of the diversity of substances listed within the category of inorganic substances.

7.5.2 Potential impact on offspring

Inorganics can have critical health impacts on human offspring. However, not all inorganics have been identified and possess the potential to impact the fetus/ offspring. On the other hand, not all inorganics have been studied to determine their fetal impact. Many organizations have in place requirements that will support pregnant women from being exposed to chemicals in the work place during pregnancy. As such, during pregnancy, the mother may be reassigned to another job without penalty in pay, benefits, of company status. When it comes to women who are pregnant and may be exposed to inorganics in the home or community, it is up to them to be conscious of any potential exposure that may impact the fetus. Avoidance of these exposures are oftentimes discussed with the mother by their physician (family physician or gynecologist).

The ability of mother to provide the necessary nutrients and oxygen for her baby is a critical factor for fetal health and survival. The inability to provide the adequate amount of nutrients to the fetus can lead to fetal malnutrition. The fetus may respond and adapt to undernutrition and permanently altering the structure and function of the body. On the other hand, maternal overnutrition can have long-lasting and negative effects on the health of the offspring. There is evidence that maternal nutrition can induce epigenetic modifications of the fetal genome.[12] Therefore, during pregnancy, it is paramount that the mother avoid all possible exposures to inorganic substances.

Little information is known about the health impact on the fetus and chemical transfer for many inorganics that are in use today. There are a few chemicals that have extensive information on health effects published. One such substance that has been extensively studied is lead. There are other inorganic substances that have been studied, and the realization is that more knowledge would be helpful. Nonetheless, the knowledge known should be used for protecting the unborn fetus and developing a hazard prevention strategy for pregnant women. Inorganic arsenic is among this category of inorganic substances.

Lead can be passed from the mother to the unborn baby. If a woman has been exposed to lead and has lead stored in her bones, lead can be released into the blood during pregnancy. If a woman is exposed to lead during pregnancy, the developing fetus can be exposed. Elevated levels of lead while pregnant

- Can increase the risk of miscarriages
- Cause the baby to be born early of too small
- Damage the baby's brain, kidneys, and nervous system
- Cause learning behavior challenges.

Lead in maternal blood can cross the placenta and result in elevated blood lead levels in newborns, potentially producing negative effects on neurocognitive function.[2] All potential exposure to lead during pregnancy must be avoided to eliminate the risk of injury to the unborn baby.

When considering arsenic, there is a recognition that more that would be helpful in exposure determination. There is a lack of data cataloging the exposure and toxic effect of inorganic arsenic during the early stages of human development. Therefore, the transfer of arsenic to the fetus and suckling infants was studied in a native Andean population. This population was living in section of Argentina where the drinking water contained elevated levels of arsenic. This study showed that arsenic is easily transferred to the fetus during the late gestation.[10] Although data are limited, the aforementioned study does provide enough information on the substance that should be used to avoid fetal exposure.

The Physicians for Social Responsibility lists some chemicals that can pose a hazard to the fetus including pesticides (impaired fetal growth) and solvents (fetal loss and miscarriages) warning of the potential health impact that the fetus may suffer.[19]

7.6 TREATMENT OF RESULTANT ILLNESS

The class of inorganic chemicals represents a variety of different chemicals of more than 100,000 substances. Therefore, treating illnesses that may require different treatment methods once exposed and at the onset of a disease is confirmed. Exposure may require treatment for health issues such as minor dermatitis to lung disease to even death. Because of the type of injury and illness that can be sustained, a physician may approach injury care in various ways. The type of health care provided is dependent on factors such as the following:

- The inorganic chemical to which exposure occurred
- The route of exposure
- The amount of substances an individual is exposed to
- The length of time for exposure
- The frequency of exposure
- Predisposed conditions of the individual exposed
- Environmental conditions in which exposure occurred

The attending physician must take into considerations all or the above as well as other factors in determining the treatment regime for exposure to inorganic substances.

REFERENCES

1. Ahmad, M.Z., Abdel-Wahab, B.A., Alam, A., Zafar, S., Ahmad, J., Ahmad, F., Midoux, P., Pichon, C., & Akhter, S., (2016). Toxicity of Inorganic Nanoparticles Used in Targeted Drug Delivery and Other Biomedical

Application: An Updated Account on Concern of Biomedical Nanotoxicology. *Journal of Nanoscience and Nanotechnology, 16* (8), 7873–7879. https://doi. org/10.1166/jnn.2016.13032.

2. Carrel, M., Zahrieh, D., Young, S.G., Oleson, J., Ryckman, K.K., Wels, B., Simmons, D.L., & Saftlas, A. (2017). High prevalence of elevated blood lead levels in both rural and urban Iowa newborns: Spatial patterns and area-level covariates. *PLoS One, 12*(5), e0177930. https://doi.org/10.1371/journal. pone.0177930.

3. Villars, Pierre Cenzual, Karin Gladyshevskii, Roman. (2013). Handbook of Inorganic Substances. De Gruyter. Retrieved from https://app.knovel. com/hotlink/toc/id:kpHIS0000B/handbook-inorganic-substances/ handbook-inorganic-substances.

4. Agency for Toxic Substances and Disease Registry (ASTDR), Toxic Substance Portal, Inorganic substances | Chemical Classifications | Toxic Substance Portal | ATSDR (cdc.gov).

5. Sekhon, B.S., & Gandhi, L. (2006). Medicinal uses of inorganic compounds -1. *Resonance. Journal of Science Education*, 11(4), 75–89. https://doi. org/10.1007/BF02835732.

6. Zoroddu, M.A., Aaseth, J., Crisponi, G., Medici, S., Peana, M., & Nurchi, V.M. (2019). The essential metals for humans: a brief overview, *Journal of Inorganic Biochemistry, 195*, 120–129.

7. Bunge, M.L. (1985). Chemical hazards in the household: What every commu-nity health nurse should know. *Journal of Community Health Nursing, 2*(1), 31–40. http://www.jstor.org/stable/3427187.

8. Schlesinger, R.B. (2007). The health impact of common inorganic components of fine particulate matter (PM2.5) in ambient air: A critical review. *Inhalation Toxicology, 19*, 811–832. https://doi.org/10.1080/08958370701402382.

9. Park, J. D., & Zheng, W. (2012). Human exposure and health effects of inorganic and elemental mercury. *Journal of preventive medicine and pub-lic health = Yebang Uihakhoe chi, 45*(6), 344–352. https://doi.org/10.3961/ jpmph.2012.45.6.344

10. Concha, G., Vogler, G., Lezcano, D., Nermell, B., & Vahter, M. (1998). Exposure to inorganic arsenic metabolites during early human development. *Toxicological Sciences, 44*(2), 185–190.

11. Kitamura, H., Ichinose, S., Hosoya, T., Ando, T., Ikushima. S., Oritsu, M., & Takemura, T. (2007). Inhalation of inorganic particles as a risk factor for idiopathic pulmonary fibrosis – elemental microanalysis of pulmonary lymph nodes obtained at autopsy cases. *Pathology Research Practice, 203*(8), 575–85. https://doi.org/10.1016/j.prp.2007.04.008.

12. Martin-Gronert M.S., & Ozanne, S.E. (2006). Maternal nutrition during pregnancy and health of the offspring. *Biochemical Society Transactions, 34*(5), 779–782. https://doi.org/10.1042/BST0340779.

13. Gorguner, M., & Akgun, M. (2010). Acute inhalation injury. *The Eurasian Journal of Medicine, 42*(1), 28–35. https://doi.org/10.5152/eajm.2010.09.

14. U.S. Department of Health and Human Services (2022). Childhood Lead Poisoning Prevention – Pregnant Women. National Center for Environmental Health, Centers for Disease Control and Prevention, U.S. Department of Health and Human Services. https://www.cdc.gov/nceh/lead/prevention/ pregnant.htm.

15. U.S. Environmental Protection Agency (2022). Exposure Assessment Tools by Chemical Classes - Inorganics and Fibers. EPA ExpoBox, U.S. Environmental Protection Agency. https://www.epa.gov/expobox/exposure-assessment-tools-chemical-classes-inorganics-and-fibers.

16. Malmquist, S. & Prescott, K. (2022). Human Biology. (pp. 379–398). University of Minnesota Open Library. https://open.lib.umn.edu/humanbiology/.

17. New Jersey Department of Health and Senior Services (2002). Hazardous Substance Fact Sheet – Ammonium Thiocyanate. New Jersey Department of Health and Senior Services. https://nj.gov/health/eoh/rtkweb/documents/fs/0119.pdf.

18. Physicians for Social Responsibility (2018). Prenatal Exposure to Toxic Chemicals. Physicians for Social Responsibility. https://psr.org/wp-content/uploads/2018/05/prenatal-exposure-to-chemicals.pdf.

19. National Capital Poison Center (2022). Poison Statistics: National Data 2020. National Capital Poison Center. https://www.poison.org/poison-statistics-national.

Chapter 8

Chemical carcinogens

8.1 INTRODUCTION

Carcinogens are agents that can cause cancer in humans. They can be divided into three major categories: chemical carcinogens (including those from biological sources), physical carcinogens, and oncogenic (cancer-causing) viruses. Most carcinogens, singly or in combination, produce cancer by interacting with **deoxyribonucleic acid** (DNA) in cells and thereby interfering with normal cellular function. This ultimately results in the formation of a tumor (an abnormal tissue growth) that can spread (metastasize) from its site of origin and invade and cause dysfunction of other tissues, culminating in organ failure and death. Several of the carcinogen classifications deal with groups of substances: aniline and homologs, chromates, dintrotoluenes, arsenic and inorganic arsenic compounds, beryllium and beryllium compounds, cadmium compounds, nickel compounds, and crystalline forms of silica. There are also substances of variable or unclear chemical makeup that are considered carcinogens, coal tar pitch volatiles, coke oven emissions, diesel exhaust, and environmental tobacco smoke.

The two primary mechanisms by which carcinogens initiate the formation of such tumors are via alterations in DNA that encourage cell division and that prevent cells from being able to self-destruct when stimulated by normal triggers, such as DNA damage or cellular injury (a process known as apoptosis). There also exist carcinogens that induce cancer through nongenotoxic mechanisms, such as immunosuppression and induction of tissue-specific inflammation.

More than 400 chemical agents have been listed as carcinogenic, probably carcinogenic, or possibly carcinogenic by the International Agency for Research on Cancer (IARC), a branch of the World Health Organization that monitors cancer occurrence worldwide and performs epidemiological and laboratory investigations to understand the causes of cancer. Among the carcinogenic substances listed by IARC are a variety of chemical effluents from industry and environmental pollutants from automobiles, residences, and factories. One such example is acrylamide, which is considered a probable carcinogen in humans and is produced because of industrial

DOI: 10.1201/9781003220114-8

processes and cooking certain foods at high temperatures. It can be released into the environment through its application in wastewater treatment and its use in grout and soil-stabilizer products. Other examples of chemical carcinogens include nitrosamines and polycyclic aromatic hydrocarbons, which are found in tobacco smoke and are associated with the development of lung cancer.

Physical carcinogens include ultraviolet rays from sunlight and ionizing radiation from X-rays and from radioactive materials in industry and in the general environment. Repeated local injury (e.g., wounding) or recurring irritation (e.g., chronic inflammation) to a part of the body are other examples of potential physical carcinogens.

A number of viruses are suspected of causing cancer in animals, including humans, and are frequently referred to as oncogenic viruses. Examples include human papillomaviruses (HPV), the Epstein-Barr virus (EBV), and the hepatitis B virus (HBV), all of which have genomes made up of DNA. Human T-cell leukemia virus type I (HTLV-I), which is a retrovirus (a type of RNA virus), is linked to tumor formation in humans.

There are some cancers that are heritable in the sense that a predisposition exists, awaiting a convergence of carcinogenic influences for cancer to manifest itself. The identification and timely elimination of carcinogens can reduce the incidence of cancer. Some of the potential carcinogens listed in this index may be re-evaluated by National Institute for Occupational Safety and Health (NIOSH) as new data become available and the NIOSH recommendations on these carcinogens either as to their status as a potential occupational carcinogen or as to the appropriate recommended exposure limit may change.

8.1.1 Mechanism of cancer formation

When chemical carcinogens are internalized by cells, they are often metabolized, and the resulting metabolic products are either excreted or retained by the cell. Inside the cell, carcinogens or their metabolic products can either directly or indirectly affect the regulation and expression of genes involved in cell-cycle control, DNA repair, cell differentiation, or apoptosis. Some carcinogens act by genotoxic mechanisms, such as forming DNA adducts or inducing chromosome breakage, fusion, deletion, missegregation, and non-disjunction. For example, carcinogenic ions or compounds of nickel, arsenic, and cadmium can induce structural and numerical chromosome aberrations. Others act by nongenotoxic mechanisms such as induction of inflammation, immunosuppression, formation of reactive oxygen species, activation of receptors such as arylhydrocarbon receptor or estrogen receptor and epigenetic silencing. Together, these genotoxic and non-genotoxic mechanisms can alter signal-transduction pathways that finally result in hypermutability, genomic instability, loss of proliferation control, and resistance to apoptosis – some of the characteristic features of cancer cells.[5]

8.1.2 Common carcinogens

8.1.2.1 Asbestos

Asbestos is a known carcinogen and is a group of six naturally occurring minerals made up of heat-resistant fibers. A list of documented carcinogens is shown in Table 8.1. Asbestos was used in thousands of US consumer products before the dangers of asbestos were known. Asbestos causes mesothelioma, lung cancer, and other cancers. Asbestos is regulated in the United States. Asbestos was widely used in construction as an effective insulator, and it can be added to cloth, paper, cement, plastic, and other materials to make them stronger. But when asbestos dust is inhaled or ingested, the fibers can become permanently trapped in the body. Over decades, trapped asbestos fibers can cause inflammation, scarring, and eventually genetic damage. There is no amount of asbestos exposure that is safe, but workers who are repeatedly exposed to asbestos or exposed to increased concentration has increased probability of worse adverse effects. Fibers are inhaled when they become airborne. It is important to avoid disturbing products that may contain asbestos.

Mesothelioma is a rare type of cancer that occurs in the thin protective layer of tissue lining important internal organs such as the heart and lungs. Mesothelioma accounts for less than 1% of all cancers in the United States. It is caused by exposure to asbestos, usually at the workplace, and can take as long as 10 – 50 years to develop after the first exposure to asbestos. When individuals are exposed to asbestos, the fibers work their way through the lungs and into the lining surrounding internal organs eventually causing cancer. This membrane surrounding these vital organs is called the mesothelium, thus the name mesothelioma.[6] High rates of mesothelioma occur in people who mine asbestos, manufacture asbestos products, work with asbestos products, live with asbestos workers, or work in buildings or homes containing asbestos.

8.1.2.2 Benzene

Benzene is a chemical that is a colorless or light-yellow liquid at room temperature. It has a sweet odor and is highly flammable. Benzene is an aromatic hydrocarbon that is produced by the burning of natural products. It is a component of products derived from coal and petroleum and is found in gasoline and other fuels. Benzene is among the 20 most widely used chemicals in the United States. It is used mainly as a starting material in making other chemicals, including plastics, lubricants, rubbers, dyes, detergents, drugs, and pesticides.

A major source of benzene exposure is tobacco smoke. Benzene is a known carcinogen. Exposures to benzene in the long term may affect bone marrow and blood production and cause leukemias. Increased incidence

Table 8.1 Occupational carcinogen list[9]

A	C	D
Acetaldehyde	Cadmium dust and fume	DDT (dichlorodiphenyltrichloroethane)
2-Acetylaminofluorene	Captafol	Di-2-ethylhexyl phthalate (DEHP)
Acrylamide	Captan	2,4-Diaminoanisoleo
Acrylonitrile	Carbon black (exceeding 0.1% PAHs)	o-Dianisidine-based dyes
Aldrin	Carbon tetrachloride	1,2-Dibromo-3-chloropropane (DBCP)
4-Aminodiphenyl	Chlordane	Dichloroacetylene
Amitrole	Chlorinated camphene	p-Dichlorobenzene
Aniline and homologs	Chlorodiphenyl (42% chlorine); class polychlorinated biphenyls	3,3′-Dichlorobenzidine
o-Anisidine	Chlorodiphenyl (54% chlorine); class polychlorinated biphenyls	Dichloroethyl ether
Arsenic and inorganic arsenic compounds	Chloroform	1,3-Dichloropropene
Arsine	Chloromethyl methyl ether	Dieldrin
Asbestos	bis(Chloromethyl) ether	Diesel exhaust
Asphalt fume	B-Chloroprene	Diglycidyl ether (DGE); class, glycidyl ethers
	Chromium, hexavalent [Cr(VI)]	4-Dimethylaminoazobenzene
B	Chromyl chloride; class, chromium hexavalent	Dimethyl carbamoyl chloride
Benzene	Chrysene	1,1-Dimethylhydrazine; class, hydrazines
Benzidine	Coal tar pitch volatiles; class, coal tar products	Dimethyl sulfate
Benzidine-based dyes	Coke oven emissions	Dinitrotoluene
Beryllium		Dioxane
Butadiene		
tert-Butyl chromate; class, chromium hexavalent		
E	**F**	**H**
Environmental tobacco smoke	Formaldehyde	Heptachlor
Epichlorohydrin		Hexachlorobutadiene

(Continued)

Table 8.1 (*Continued*) Occupational carcinogen list[9]

A	C	D
Ethyl acrylate	**G**	Hexachloroethane
Ethylene dibromide	Gallium arsenide	Hexamethyl phosphoric triamide (HMPA)
Ethylene dichloride	Gasoline	Hydrazine
Ethylene oxide		
Ethyleneimine		**K**
Ethylene thiourea		Kepone
M	**P**	**T**
Malonaldehyde	Pentachloroethane; class, chloroethanes	Tremolite silicates
Methoxychlor	N-Phenyl-b-naphthylamine; class, b-naphthalene	2,3,7,8-Tetrachlorodibenzo-p-dioxin (TCDD) (dioxin)
Methyl bromide; class, monohalomethanes	Phenyl glycidyl ether; class, glycidyl ethers	1,1,2,2-Tetrachloroethane
Methyl chloride	Phenylhydrazine; class, hydrazines	Tetrachloroethylene
Methyl iodide; class, monohalomethanes	Propane sultone	Titanium dioxide
Methyl hydrazine; class, hydrazines	B-Propiolactone	o-Tolidine-based dyes
4,4′-Methylenebis(2-chloroaniline) (MBOCA)	Propylene dichloride	o-Tolidine
Methylene chloride	Propylene imine	Toluene diisocyanate (TDI)
4,4-Methylenedianiline (MDA)	Propylene oxide	Toluene diamine (TDA)
		o-Toluidine
N	**R**	p-Toluidine
α-Naphthylamine (alpha-naphthylamine)	Radon	1,1,2-Trichloroethane; class, chloroethanes
β-Naphthylamine (beta-naphthylamine)	Rosin core solder, pyrolysis products (containing formaldehyde)	Trichloroethylene
Nickel, metal, soluble, insoluble, and inorganic; class, nickel, inorganic		1,2,3-Trichloropropane
Nickel carbonyl	**S**	
Nickel sulfide roasting	Silica, crystalline cristobalite	**U**
4-Nitrobiphenyl	Silica, crystalline quartz	Uranium, insoluble compounds Uranium, soluble compounds

(*Continued*)

Table 8.1 (Continued) Occupational carcinogen list[9]

A	C	D
p-Nitrochlorobenzene	Silica, crystalline Tripoli	
2-Nitronaphthalene	Silica, crystalline tridymite	**V**
2-Nitropropane	Silica, fused	Vinyl bromide; class, vinyl halides
N-Nitrosodimethylamine	Soapstone, total dust silicates	Vinyl chloride
		Vinyl cyclohexene dioxide
		Vinylidene chloride (1,1-dichloroethylene); class, vinyl halides)
		W
		Welding fumes, total particulates
		Wood dust
		Z
		Zinc chromate; class, chromium hexavalent

of leukemia (cancer of the tissues that form white blood cells) has been observed in humans occupationally exposed to benzene.

The permissible exposure limit (PEL) level of benzene is 1 part per million (ppm) in air for an 8 hour average. The short-term exposure limit (STEL) is 5 ppm. Benzene can also be absorbed through the skin and into the bloodstream. Workers in industries that manufacture or use benzene may be exposed to the highest levels of benzene.

8.1.2.3 Hexavalent chromium

Hexavalent chromium (Cr (VI)) compounds are a large group of chemicals with varying properties, uses, and workplace exposures. Chromium compounds, such as hexavalent chromium, are widely used in electroplating, stainless steel production, leather tanning, textile manufacturing, and wood preservation. The United States is one of the world's leading producers of chromium compounds. Hexavalent chromium compounds are considered carcinogenic to workers. The risk of developing lung, nasal, and sinus cancer increases with the amount of hexavalent chromium inhaled and the length of time the worker is exposed. Certain hexavalent chromium compounds produced lung cancer in animals that had the compounds placed directly in their lungs.

8.1.2.4 Formaldehyde

Formaldehyde is a colorless, flammable, strong-smelling chemical that is used in building materials and to produce many household products. Formaldehyde is commonly used as an industrial fungicide, germicide, and disinfectant, and as a preservative in mortuaries and medical laboratories. When dissolved in water, it is called formalin, which is commonly used as an industrial disinfectant, and as a preservative in funeral homes and medical labs. Workers in industries that make formaldehyde or formaldehyde-containing products, lab technicians, some health care professionals, and funeral home employees may be exposed to higher levels of formaldehyde. Exposure occurs predominately by inhaling formaldehyde gas or vapor from the air or by absorbing liquids containing formaldehyde through the skin. The US Occupational Safety and Health Administration (OSHA) has established limits for formaldehyde that workers can be exposed at work at 0.75 ppm on average of an 8 hours of workday. Studies of workers exposed to high levels of formaldehyde, such as industrial workers and embalmers, have found that formaldehyde causes myeloid leukemia and rare cancers, including cancers of the paranasal sinuses, nasal cavity, and nasopharynx.[7]

8.1.2.5 Silica

Silica is an oxide of silicon namely silicon dioxide and is generally colorless to white and insoluble in water. Respirable crystalline silica is generated in work activities that involve cutting, sawing, grinding, drilling, and crushing stone, rock, concrete, brick, block, and mortar. In addition, activities such as abrasive blasting with sand, sawing brick or concrete, sanding or drilling into concrete walls, grinding mortar, manufacturing brick, concrete blocks, stone countertops, or ceramic products, and cutting or crushing stone will lead to exposure to respirable crystalline silica dust. Industrial sand used in certain operations, such as foundry work and hydraulic fracturing (fracking), is also a source of respirable crystalline silica exposure. About 2.3 million people in the United States are exposed to silica at work. Workers who inhale these very small silica particles are at increased risk of developing lung cancer or silicosis, an incurable lung disease that can lead to disability and death.[8]

8.1.2.6 Vinyl chloride

Vinyl chloride is a colorless, flammable gas that evaporates very quickly. It is used to make polyvinyl chloride (PVC) pipes, wire coatings, vehicle upholstery, and plastic kitchenware. Workers at facilities where vinyl chloride is produced or used may be exposed primarily through inhalation. Vinyl chloride is also produced as a combustion product in tobacco smoke. Vinyl chloride exposure is associated with an increased risk of hepatic angiosarcoma, a rare form of liver cancer, brain and lung cancers, lymphoma, and leukemia.

Table 8.1 lists chemicals that are listed as carcinogens that are encountered in the occupational environment.

Identifying carcinogens is complicated. Fortunately, there are several organizations that evaluate the available information according to specific criteria.

The most authoritative lists of carcinogens are published by the following:

- International Agency for Research on Cancer (IARC), an agency of the World Health Organization
- American Conference of Governmental Industrial Hygienists (ACGIH), an independent US organization
- US National Toxicology Program (NTP), a US interagency program

8.1.2.7 IARC

IARC classifies each agent or exposure into one of five groups according to the strength of scientific evidence for carcinogenicity, as follows:

- Group 1 – Carcinogenic to humans
- Group 2A – Probably carcinogenic to humans
- Group 2B – Possibly carcinogenic to humans
- Group 3 – Not classifiable as to its carcinogenicity to humans
- Group 4 – Probably not carcinogenic to humans

8.1.2.8 ACGIH

ACGIH assigns chemicals or agents to one of the following five categories:

- A1– Confirmed human carcinogen
- A2 – Suspected human carcinogen
- A3 – Confirmed animal carcinogen with unknown relevance to humans
- A4 – Not classifiable as a human carcinogen
- A5 – Not suspected as a human carcinogen

Carcinogens identified by ACGIH are listed in ACGIH's TLVs® and BEIs® booklet, which is published annually. See the ACGIH website for more information.

8.1.2.9 NTP

Every 2 years, NTP publishes a list of agents that they have evaluated and assigned to one of two categories:

- Known to be human carcinogens
- Reasonably anticipated to be human carcinogens

8.2 EXPOSURE PATHWAYS

Individuals are constantly exposed to varying amounts of chemicals that have been shown to have carcinogenic or mutagenic properties. Exposure can occur exogenously when these agents are present in food, air, or water and also endogenously when they are products of metabolism or pathophysiologic states such as inflammation. It has been estimated that exposure to environmental chemical carcinogens may contribute significantly to the causation of a sizable fraction, a majority, of human cancers, when exposures are related to "lifestyle" factors such as diet and tobacco use. There are studies establishing a causal link between environmental exposures and increased cancer risks.[2]

Some chemicals are known to cause cancer in animals, but they have not been proven to cause cancer in humans. These chemicals are anticipated to cause cancer in humans and are sometimes called possible human carcinogens. Based on the rationale and methods described in Environmental Protection Agency (EPA) 2003 draft carcinogen risk assessment guidelines, EPA conducts a qualitative weight-of-evidence evaluation of human and animal toxicity studies of a substance. EPA provides weight-of-evidence narratives and presents the following descriptors to describe the carcinogenicity of a given substance:

- Carcinogenic to humans
- Likely to be carcinogenic to humans
- Suggestive evidence for carcinogenic potential
- Inadequate evidence to assess carcinogenic potential
- Not likely to be carcinogenic to humans

Earlier EPA guidelines (1986) used a slightly different cancer classification scheme, which is still in place for many substances. Under that scheme, potential carcinogens are classified as follows:

A Human carcinogen (sufficient human data)
  B1 Probable human carcinogen (limited human data, sufficient animal data)
  B2 Probable human carcinogen (inadequate human data, sufficient animal data)
C Possible human carcinogen (inadequate or no human data, sufficient animal data)
D Not classifiable as to human carcinogenicity (inadequate or no human and animal data)
E Evidence of non-carcinogenicity in humans (adequate human and animal data)

8.2.1 Inhalation exposures and impact

Inhalation of carcinogens occurs by absorption of the carcinogen through the respiratory tract. Once the carcinogen has entered the respiratory tract, it is absorbed into the bloodstream for distribution throughout the body. Carcinogens can be inhaled in the form of vapors, fumes, mists, aerosols, and fine dust. Worker's exposure to certain chemicals that are carcinogenic through inhalation. Gases and particulates liberated from burning chemical materials often contain toxic, reactive, and carcinogenic chemicals that can be inhaled.

8.2.2 Dermal exposures and impact

The skin is the body's largest organ accounting for more than 10% of total body mass. It is a very complex and dynamic organ composed of an outer epidermis and inner dermis with functions well beyond that of just a barrier to the external environment. Dermal absorption depends largely on the barrier function of the stratum corneum, the outermost superficial layer of the epidermis, and is modulated by factors such as skin integrity, hydration, density of hair follicles and sebaceous glands, thickness at the site of exposure, physiochemical properties of the substance, chemical exposure concentration, and duration of exposure. Some chemicals can enter the body through the skin and damage vital organs or groups of organs. Absorbing workplace chemicals through the skin can cause diseases and disorders that include occupational skin diseases. There are many workers in the United States from a variety of occupational industries and sectors, who are potentially exposed to chemicals that can be absorbed through the skin. Occupational skin exposures can result in numerous diseases that can adversely affect an individual's health and capacity to perform at work. In general, there are three types of chemical-skin interactions of concern: direct skin effects, immune-mediated skin effects, and systemic effects. While hundreds of chemicals (metals, epoxy and acrylic resins, rubber additives, and chemical intermediates) present in nearly every industry have been identified to cause direct and immune-mediated effects such as contact dermatitis or urticaria, less is known about the number and types of chemicals contributing to systemic effects. To raise awareness, skin notation assignments communicate the potential for dermal absorption. Studies have suggested that exposure to complex mixtures, excessive hand washing, use of hand sanitizers, high frequency of wet work, and environmental or other factors may enhance penetration and stimulate other biological responses altering the outcomes of dermal chemical exposure. Understanding the hazards of dermal exposure is essential for the proper implementation of protective measures to ensure worker safety and health.[3]

Dermal exposure to chemicals in the work environment may occur through direct contact with contaminated surfaces, deposition of aerosols, immersion, or splashes and can often occur without being noticed by the worker. This is particularly true for non-volatile chemicals, which remain on work surfaces for long periods of time. Prolonged exposures may result from contamination of clothing or permeation of chemicals through gloves, potentially resulting in enhanced absorption secondary to occlusion. Therefore, it is critical for workers to understand the significance of dermal exposure and what measures to take for prevention.[3]

There are currently no occupational exposure limits (OELs) set for dermal exposures; however, chemicals with risk associated with dermal penetration are given a skin notation assignment (S) as a guidance to warn against potential for increased risk of systemic toxicity because of dermal penetration in additional to inhalation exposure.[3]

8.2.3 Ingestion exposure and impact

Ingestion of chemical carcinogens can occur through the consumption of contaminated food, water, and other liquids. Food can contain chemical residues because of the following:

- Intentional application (e.g., pesticide use)
- Deposition of particulate matter onto edible produce (e.g., from atmospheric pollutants)
- Biotic uptake and accumulation from contaminated soil or water (e.g., irrigation water, uptake of contaminants by fish or livestock).

Ingestion exposure can also occur through the intentional or inadvertent non-dietary ingestion of soil, dust, or chemical residues on surfaces or objects that are contacted via hand-to-mouth or object-to-mouth activity.

Food-borne carcinogens constitute the primary source of ingested carcinogens.

8.3 EXPOSURE CONTROL METHODS

There are international standards for the protection of workers from risks related to exposure to carcinogens and mutagens at work. These are summarized in Table 8.2.[4]

8.3.1 Engineering controls

Engineering controls are methods used to protect workers by eliminating hazardous conditions or by placing a barrier between the worker and the hazard. These could be local exhaust ventilation to capture and remove airborne emissions or machine guards to shield the worker. Engineering

Table 8.2 Control methods

Control methods
Minimize the quantity of carcinogen used
Minimize the number of workers exposed
Design work to minimize the release of carcinogens
Use local ventilation to reduce exposure
Use appropriate procedures to detect accidental releases from work processes
Apply of suitable work control procedures and methods to reduce exposure
Provide individual protection measures such as personal protective equipment
Provide information for workers
Provide demarcation of risk areas and use of adequate warning signs
Provide emergency plans for circumstances where high exposure may occur
Provide means for safe storage, handling, and transportation
Provide means for safe waste disposal

controls can be highly effective in protecting workers. A disadvantage of engineering controls is the higher cost of implementing engineering controls.

Laboratory workers can protect themselves from exposure via inhalation through proper use of a functioning fume hood, use of respirators when a fume hood is not available, avoiding bench top use of carcinogenic chemicals, ensuring chemical containers are kept tightly capped, and ensuring all chemical spills are promptly cleaned up.

8.3.2 Administrative controls

Administrative controls are also known as best work practices. These include changes in work procedures such as written safety policies, procedures, rules, supervision, schedules, and training with the goal of reducing the duration, frequency, and severity of exposure to hazardous chemicals or situations. Administrative controls are targeted at changing the behavior of the worker and do not remove the actual hazard. Administrative controls are more effective than personal protective equipment (PPE) because they involve some form of planning and avoidance, while PPE serves only as a final barrier between the hazard and worker. Administrative controls are recommended when hazards cannot be removed or changed, and engineering controls are not feasible.

8.3.3 Protective equipment and effectiveness

When engineering, work practice, and administrative controls are not feasible or do not provide sufficient protection, employers must provide PPE to their workers and ensure its proper use. Employers are also required to train each worker required to use PPE. PPE is equipment worn to minimize exposure to chemical hazards that cause serious workplace cancers. PPE may include items such as gloves, safety glasses and shoes, earplugs or muffs, hard hats, respirators, coveralls, vests, and full-body suits.

8.4 POTENTIAL IMPACT ON EXPOSED INDIVIDUAL

There are several impacts on workers exposed to chemical carcinogens. These impacts are summarized in Table 8.3[1]:

Table 8.3 Worker exposure impacts

Chemicals or exposure	Cancer site
Arsenic and inorganic arsenic compounds	Bladder
Benzidine	
Tobacco smoke	
ortho-Toluidine	
Benzene	Blood (leukemia, lymphoma)
Ionizing radiation	
1,3-Butadiene	
Alcohol	Colon
Tobacco smoke	
Arsenic	Kidney
Cadmium and cadmium compounds	
Trichloroethylene (TCE)	
Alcohol	Liver
TCE	
Vinyl chloride	
Arsenic and inorganic arsenic compounds	Lung
Asbestos (all forms)	
Beryllium and beryllium compounds	
Cadmium and cadmium compounds	
Chromium (hexavalent) compounds	
Coke oven emissions	
Radon-222 and its decay products	
Tobacco smoke	
Alcohol	Oral cavity and pharynx
Betel nut use	
Tobacco use (smoking and smokeless)	
Arsenic and inorganic arsenic compounds	Skin
Coal tars	

8.5 POTENTIAL IMPACT ON OFFSPRING

Perinatal exposure to carcinogens may contribute to cancer susceptibility. This has been supported by studies that stipulate that exposure to carcinogens in utero increases the risk of the offspring developing cancer post-delivery.

8.6 TREATMENT OF RESULTANT ILLNESS

Symptoms of exposure to carcinogens through inhalation include eye, nose, and throat irritation, coughing, difficulty in breathing, headache, dizziness, confusion, and collapse. If any of these symptoms are noted, leave the area immediately, and get fresh air. Seek medical attention if symptoms persist, and complete Injury/Illness Report.

REFERENCES

1. Burdick, G. (2021) Carcinogens in the Workplace: Considerations for EHS Chemicals, ESH Management, Injuries, and Illness https://ehsdailyadvisor.blr.com/2021/03/carcinogens-in-the-workplace-considerations-for-ehs/
2. Wogan, G.N., Hecht, S.S., Felton, J.S., Conney, A.H., & Loeb, L.A. (2004). Environmental and Chemical Carcinogenesis. *In Seminars in Cancer Biology*, 14(6), 473–486. Academic Press.
3. Anderson, S. E., & Meade, B. J. (2014). Potential health effects associated with dermal exposure to occupational chemicals. *Environmental Health Insights*, 8, EHI-S15258. PMID: 25574139; PMCID: PMC4270264.
4. Cherrie, J.W. (2009). Reducing occupational exposure to chemical carcinogens. *Occupational Medicine*, 59(2), 96–100.
5. Arya A., Arya S. and Arya M. (2011). Chemical Carcinogen and Cancer Risk: An overview. *Journal of Chemical and Pharmaceutical Research*, 3(5), 621–631.
6. Roberts, C. (2021). Why is Asbestos so dangerous? Veterans Asbestos Alliance https://www.veteransasbestosalliance.org/post/why-is-asbestos-so-dangerous.
7. National Cancer Institute (2019). Formaldehyde. National Cancer Institute. National Institutes of Health, U.S. Department of Health and Human Services. https://www.cancer.gov/about-cancer/causes-prevention/risk/substances/formaldehyde.
8. Occupational Health and Safety Amendment (Crystalline Silica) Regulations 2021 S.R. No. 137/2021, https://content.legislation.vic.gov.au/sites/default/files/2021-11/21-137sra%20authorised_0.pdf.
9. https://www.cdc.gov/niosh/topics/cancer/npotocca.html.

Chapter 9

Mutagens

9.1 INTRODUCTION

A mutagen is defined as an agent that causes irreversible and heritable changes (mutations) in the cellular genetic material, deoxyribonucleic acid (DNA). Mutations are produced if the damaged nucleotides are incorporated into DNA incorrectly. Most mutagens are endogenous agents that are present in the cell under normal physiological conditions and include reactive oxygen species (ROS) and alkylating agents. Mutagens are also present in the environment and include sunlight and a multitude of chemicals that may be ingested in the foods that we eat or in the air that we breathe. Mutagens that promote the development of cancer are called carcinogens.

Mutagenic lesions persist when they escape detection by protective cellular DNA repair mechanisms, when mistakes occur in the repair process, or when repair mechanisms are overwhelmed by extensive damage. The biological consequences of a mutation depend on many critical factors such as the target loci, size of the mutation, timing during the cell cycle, and compounding effects of preexisting mutations. Thus, a mutagenic event occurring in a non-functional area of DNA will have no effect (silent mutation), whereas a similar change in an actively transcribed region may profoundly affect gene expression and phenotype or even lead to cell death (lethal mutation).

The influence of mutations in human health is underscored by several human disease states caused by mutations that disrupt regulatory regions or gene coding sequences, resulting in altered gene expression and protein function. There are many significant pathways to the development of carcinogenesis such as mutations in genes that promote or inhibit growth and cellular replication or code for components of DNA repair. Significant exposure to mutagenic compounds can occur through food and water, as well as through environmental and occupational sources. Most developed countries have instituted regulatory protocols governing the introduction of new foods and food-associated chemicals before they are accepted into the marketplace to minimize exposure to mutagens. Mutagens are factors capable of affecting the structural integrity of the genome. The number of

DOI: 10.1201/9781003220114-9

harmful mutagens we are continually exposed to, and interacting with, in our life, starting with fertilization and ending with death, is innumerable. However, due to the many lines of defense against the harmful effects of these factors, represented by the many anti-mutation mechanisms of the genome and proteome, only a small percentage of mutational events can reach a threshold level beyond which genetic defects can develop.

9.2 CLASSIFICATION OF MUTAGENS

(A) According to their nature, mutagens are classified into three main categories:
- Chemical mutagens: these compounds are innumerable in the environment and include, for example, organic compounds, asbestos, insecticides, herbicides, heavy metals, etc.
- Physical mutagens: these include particulate radiations like X-ray, alpha particles, UV waves at 2,800 A° wavelength, solar radiation, and thermal and mechanical agitation of nucleic acids[1,2].
- Biological mutagens: these include living microorganisms like some viruses: cytomegalovirus, rubella virus, and herpes virus.

(B) According to their pathogenetic effects, mutagens are classified into four main categories:
- Non-specific mutagens: these unclassified mutagens cause DNA damage and non-functioning of the DNA repair pathway.
- Carcinogens: these are mutagens that induce malignant transformations in affected cells.
- Clastogens: these are mutagenic agents that can induce chromosome breaks in affected cells.
- Teratogens: these are mutagens that cause congenital malformations in exposed fetuses.

The pathogenetic effects of mutagens and the resulting pathological alterations follow disturbed functions of mutated genes. Many types of mutagens can induce all types of mutagenic damage to the cell. Particulate and non-particulate radiations, for example, are effective carcinogens capable of causing malignant transformation of affected cells. They can also cause chromosomal gaps and breaks. In addition, teratogenic effects in developing fetuses following exposure of pregnant women to radiation are a recognized hazard of overexposure to radiation during pregnancy. This global mutagenic effect is also shared by many other types of mutagens like mutagenic viruses and chemicals that can induce malignant transformation, chromosomal breaks, and teratogenic malformations in exposed subjects[2].

9.3 FACTORS AFFECTING ACTIONS OF MUTAGENS

The pathological effects of mutagens are determined by many different factors. Each of these factors has its relative mutagenic potential depending on the circumstances of its action. Congenital malformations caused by teratogens, for instance, depend largely on the timing of exposure of the developing embryo or growing fetus to their effects. Exposure to teratogens before or after periods of embryonic or fetal growth sensitive to their specific teratogenic effects might result in no or minimal harm. The dose of exposure, whether of a chemical or radiological nature, is also an important factor in determining the resultant mutagenic effects. This dose-threshold effect might act in different synergistic ways. A larger exposure dose will quantitatively cause more damage to wider portions of the genetic material, thus resulting in widespread mutations of the genome. It might also cause damage to genes responsible for the repair of mutated DNA, thus preventing the repair of mutated genes. In addition, direct non-genetic destructive effects of cellular compartments, of blood supply to cells, or of extracellular environment of affected cells will both hasten and increase the resulting damage. The genetic constitution, or genetic background, of exposed subjects plays a critical role in determining the sensitivity to mutagenic factors as well as the extent of damage following exposure to their effects. The outstanding examples in this respect are the chromosome breakage syndromes, sometimes referred to as DNA or genetic instability syndromes. These diseases include xeroderma pigmentosum, ataxia telangiectasia, Nijmegen syndrome, Bloom syndrome, Fanconi anemia, and some other similar diseases. Subjects affected by these genetic disorders have deficient and/or defective DNA repair systems necessary for repairing mutations arising during DNA replication or following exposure to mutagens. Exposure to mutagenic factors in doses that have no effects in normal subjects causes drastic genetic alterations in these patients due to their defective genetic constitution[2].

9.4 COMMON TYPES OF MUTAGENS

There are many types of mutagens that are of concern that can have a major impact on the human system. Many of these mutagens are discussed in Sections 9.4.1–9.4.4.

9.4.1 Base analogues

These agents possess structural similarities to the bases, that is, purines and pyrimidines. The most common base analogues which are considered as chemical mutagens are – 5-Bromouracil and aminopurine. Due to structural similarities, these agents with the DNA bases, base analogues get incorporated into the DNA structure during the process of replication.

- 5-Bromouracil is a brominated derivative of uracil that acts as an antimetabolite or base analogue, substituting for thymine in DNA, and can induce DNA mutation in the same way as 2-aminopurine. It is used mainly as an experimental mutagen, but its deoxyriboside derivative (5-bromo-2-deoxy-uridine) is used to treat neoplasms.
- Aminopurine, a purine analogue of guanine and adenine, is a fluorescent molecular marker used in nucleic acid research. It most commonly pairs with thymine as an adenine analogue but can also pair with cytosine as a guanine analogue. For this reason, it is sometimes used in the laboratory for mutagenesis.

9.4.2 Intercalating agents

These are the molecules that have a hydrophobic heterocyclic ring structure and resemble the ring structure of base pairs. These agents place themselves in the DNA helix, which eventually interferes with the replication, translation, and transcription resulting in mutation, most commonly frameshift mutation. Ethidium bromide, proflavine, acridine orange, actinomycin D, and daunorubicin are some of the common intercalating agents.

- Ethidium bromide is an intercalating agent commonly used as a nucleic acid stain in molecular biology laboratories for techniques such as agarose gel electrophoresis. When exposed to ultraviolet light, it will fluoresce with an orange color, intensifying almost 20-fold after binding to DNA. Ethidium bromide is a highly toxic mutagen because it can bind with DNA. Exposure routes of ethidium bromide are inhalation, ingestion, and skin absorption. Acute exposure to ethidium bromide causes irritation of the mouth, upper respiratory tract, skin, and eyes.
- Proflavine acts by interchelating DNA (intercalation), thereby disrupting DNA synthesis and leading to high levels of mutation in the copied DNA strands. It is an acriflavine derivative, which is a disinfectant bacteriostatic against many gram-positive bacteria.
- Acridine orange is a cell-permeant nucleic acid–binding dye that emits green fluorescence when bound to dsDNA and red fluorescence when bound to ssDNA or RNA. This unique characteristic makes acridine orange useful for cell-cycle studies. Acridine orange has also been used as a lysosomal dye.
- Actinomycin D is a non-fluorescent DNA intercalator that exhibits high GC selectivity and causes distortion at its binding site. Binding of the non-fluorescent actinomycin D to nucleic acids changes the absorbance of the dye. Actinomycin D has been used for chromosome banding studies.

9.4.3 Metal ions

Ions like nickel, chromium, cobalt, cadmium, arsenic, and iron, generate ROS that cause DNA hypermethylation, thereby promoting DNA damage and hindering the DNA repair process.

Nickel: Studies have determined the mutagenic properties of nickel sulfate, nickel subsulfide, nickel hydroxide, nickel oxide (green), and nickel metal powder. Certain nickel compounds promote cell proliferation, which would convert repairable DNA lesions into non-repairable mutations.[1]

Chromium: Hexavalent chromium is a toxic form of the element chromium. Cr(VI) compounds are man-made and widely used in many different industries. Inhalation of hexavalent chromium increases the risk of developing lung cancer. Breathing in high levels of Cr(VI) can cause other effects on the respiratory system including irritation (e.g., runny nose, sneezing, coughing, itching, and a burning sensation). Repeated or prolonged exposure can cause sores to develop in the nose and result in nosebleeds. If the damage is severe, the nasal septum can develop a hole. Respiratory system sensitization can also occur, causing asthma symptoms (e.g., wheezing and shortness of breath). Cr(VI) compounds are powerful skin irritants. Contact with non-intact skin can lead to chrome ulcers. Some workers develop an allergic skin reaction, called allergic contact dermatitis. Cr(VI) compounds are eye irritants. Direct contact with chromic acid or chromate dust can cause permanent eye damage.

Cobalt: Cobalt and several of its compounds are used in various industrial applications. Workers are mainly exposed to cobalt by inhalation and varying toxic manifestations affecting the respiratory system have been reported in these industries. Diseases of the upper respiratory tract, the bronchial tree, and the lung parenchyma have been reported and reviewed. Some manifestations are associated with specific industrial settings involving exposure to a specific type of cobalt compound. It has, for instance, been shown, both epidemiologically and experimentally, that the occurrence of fibrosing alveolitis is specifically associated with the hard metal industry (hard metal disease) where exposure is to a mixture of cobalt metal and carbide particles.

Cadmium: Cadmium and its compounds are highly toxic with many industrial uses including their use in nuclear reactors. The primary and most serious adverse health effects of long-term exposure to cadmium include kidney dysfunction, lung cancer, and prostate cancer. Cadmium may cause local skin or eye irritation and can affect long-term health if inhaled or ingested. Exposure to cadmium by inhalation may occur where workers are exposed to cadmium dust from machined cadmium-containing metals, or fumes from heated cadmium compounds or surfaces such as welding or cutting materials or solders that contain cadmium. Exposure to cadmium may also target the body's cardiovascular, gastrointestinal, neurological, reproductive, and respiratory systems. Bone changes may occur because of cadmium affecting renal function.

Arsenic: Occupational exposure and arsenic poisoning may occur in persons working in industries involving the use of inorganic arsenic and its compounds, such as wood preservation, glass production, non-ferrous metal alloys, and electronic semiconductor manufacturing. Common inorganic arsenic compounds include arsenic pentoxide, arsenic trioxide, and arsenic acid. Acute exposures can cause irritation of the respiratory tract, skin, and eyes. Chronic exposure leads to distinct skin diseases (excessive formation of scaly skin, darkened patches of skin, wart formation, and skin lesions.

9.4.4 Alkylating agents

These agents induce alkyl groups in DNA resulting in DNA damage. The introduction of the alkyl groups increases ionization that results in base-pairing errors and eventually inducing gaps in the DNA strand. Some of the common alkylating agents are ethylnitrosourea (ENU), mustard gas, vinyl chloride, methylhydrazine, busulfan, carmustine, lomustine, dimethyl sulfate, temozolomide, dacarbazine, ethyl ethane sulfate, and thio-TEPA.

ENU: This is a highly potent mutagen. It is an alkylating agent and acts by transferring the ethyl group of ENU to nucleobases (usually thymine) in nucleic acids. It is known to induce point mutations, which implies that by mapping for the desired phenotype.

Mustard gas: Mustard gas, or sulfur mustard, is a chemical agent that causes severe burning of the skin, eyes, and respiratory tract. When pure, it is, in fact, both odorless and colorless. The effects of mustard gas exposure include the reddening and blistering of skin, and, if inhaled, will also cause blistering to the lining of the lungs, causing chronic impairment, or death. Sulfur mustards readily eliminate chloride ions by intramolecular nucleophilic substitution to form cyclic sulfonium ions. These very reactive intermediates tend to permanently alkylate nucleotides in DNA strands, which can prevent cellular division, leading to programmed cell death. Their alkylating capabilities make them strongly mutagenic.

Vinyl chloride: Vinyl chloride is a colorless gas that burns easily. It does not occur naturally and must be produced industrially for its commercial uses. It is used primarily to make polyvinyl chloride (PVC); PVC is used to make a variety of plastic products, including pipes, wire and cable coatings, and packaging materials. Vinyl chloride is also produced as a combustion product in tobacco smoke. Workers at work sites where vinyl chloride is produced or used may be exposed primarily through inhalation.

9.5 EXPOSURE PATHWAYS

The quantitative mutagenicity risk assessment deals with the question of how much mutational damage is likely to be produced by exposure to a given agent under particular exposure scenarios. In a dose-response assessment, the relationship between the dose of a chemical and the probability of induction of an adverse effect is defined. The component generally entails an extrapolation from the high doses administered to experimental animals or noted in some epidemiologic studies to the low exposure levels expected from human contact with the chemical in the environment

9.5.1 Inhalation exposures and impact

Inhalation of mutagens occurs by uptake of the mutagen through the respiratory tract. Once the mutagen has entered the airways, it is absorbed into the bloodstream for distribution throughout the body. Mutagens can be inhaled as fumes, vapors, mists, aerosols, and fine dust. Exposure of workers to certain chemicals that are mutagenic occurs when inhaled. Gases and particulates released from burning chemical materials often contain toxic, reactive, and mutagenic chemicals that can be inhaled.

9.5.2 Dermal exposures and impact

Some chemicals can enter the body through the skin and damage vital organs. The absorption of mutagenic chemicals at work through the skin can cause diseases and disorders including occupational skin diseases. In the United States, there are many workers in different occupations who may be exposed to mutagenic chemicals that can be absorbed through the skin. Occupational skin contact can cause numerous diseases that can affect a worker's health and performance.

Chemical-skin interactions can occur in three ways namely direct skin effects, immune-mediated skin effects, and systemic effects. Although hundreds of chemicals (metals, epoxies and acrylics, rubber additives, and chemical intermediates) found in almost every industry have been identified as causing direct and immune-mediated effects such as contact dermatitis or urticaria, less is known about the quantity and types of chemicals and contribution to systemic effects. The assignments of skin notations communicate the absorption potential of the skin are used to inform users of chemicals. Studies have shown that exposure to complex mixtures, excessive hand washing, use of hand sanitizer, frequent wet work, and environmental or other factors can enhance penetration and stimulate other biological responses that alter dermal chemical exposure outcomes. Understanding the hazards of dermal exposure is essential for the proper implementation of protective measures to ensure the safety and health of workers. Dermal exposure to mutagenic chemicals in the work environment can occur through direct contact with contaminated surfaces, aerosol deposition, immersion, or splashing and can often be unnoticed by the worker. This is especially true for non-volatile chemicals that remain on work surfaces for long periods of time. Prolonged contact can be caused by clothing contamination or chemical penetration through gloves, which can result in increased absorption because of the occlusion. Therefore, it is important that workers understand the importance of dermal exposure and what preventive measures to take. There are currently no established occupational exposure limit values (OELs) for dermal exposures; However, chemicals with a risk associated with dermal penetration are given the skin notation (S) as a guide to warn of a possible increased risk of systemic toxicity due to dermal penetration in addition to inhalation exposure.

9.6 EXPOSURE CONTROL METHODS

When chemical mutagen such as pure ethidium bromide is used, handling should be performed in a fume hood with the worker wearing full protection clothing including a lab coat, closed-toe shoes, chemical-resistant gloves, and chemical safety goggles. Gloves made of nitrile rubber are used for hand protection, and when working with high concentrations or for a prolonged period, double gloving can further reduce the risk of exposure, especially if the outer glove is replaced when contaminated. Workers should wash their hands after removing their gloves, even if the gloves are not torn or punctured, to remove any residue that may have contacted the skin. An emergency eyewash and shower should be easily accessible.

9.6.1 Elimination of hazards

Elimination of the hazard or substitution with a less hazardous chemical. Engineering controls such as the use of biological safety cabinets, isolators, or closed systems. Administrative controls such as training and education programs, the availability of safety data sheets, or established work practices, policies, and surveillance. Personal protective equipment (PPE) includes the use of protective gloves, gowns, respiratory protection, and eye protection.

9.6.2 Engineering controls

Chemicals that are mutagens should be substituted with less hazardous materials. Selection of controls should be made using a risk-based approach that considers the degree of hazard, route of exposure, and characteristics of the process that may potentially lead to exposure. Workplace engineering controls such as a fume hood or gloveboxes can be used to minimize worker exposure. Chemical fume hoods and gloveboxes provide the best protection against exposure in the laboratory. Generally, chemical fume hoods are the preferred ventilation control device in laboratories unless a glovebox is warranted. Handle mutagenic materials in a fume hood or glovebox.

Clean the work area and surfaces (e.g., workbench or work product) by wet wiping with a pre-wetted disposable cloth. HEPA vacuuming at the end of each shift.

Wash hands and face immediately after handling cadmium and before leaving the work area. Limit access to authorized and assigned personnel only.

9.6.3 Administrative controls

Administrative controls in terms of best work practices are essential to protect workers exposed to chemical mutagens. Protective equipment and environments should be accompanied by a stringent program of work practices, including operator training, and demonstrated competence, contamination reduction, and decontamination. Employee training must conform to the requirements of the Hazard Communication Standard (HCS) and other relevant OSHA requirements, such as the PPE standard [29 CFR 1910.132]. Training required under the HCS must include all employees potentially exposed to these agents.

9.6.4 Personal protective equipment and effectiveness

It is recommended workers use double gloves. Generally, nitrile or neoprene provides adequate protection against minor splashes for most chemicals. Consult glove manufacturer's chemical compatibility guides for best glove selection, or alternative glove types if needed for specific chemicals that penetrate nitrile or neoprene. SDS recommendations on glove type should be reviewed. Other PPEs include splash goggles, lab coat, clothing that leaves no exposed skin on legs or feet, chemical-resistant apron, and face shield. Closed-toe shoes that fully cover the top of the foot are recommended for foot protection.

9.6.4.1 Dermal protective PPE

PPE can be effective in providing a barrier between the skin and the constituents of concern. Options to protect the skin from exposure can include wearing long-sleeved lab or shop coats, placing coveralls over work clothes, and wearing closed-toe and closed-heel shoes. Safety glasses with side shields are a good option to protect the eyes.

9.6.4.2 Respiratory protection

In situations where there is a need to wear respiratory protection a half-face APR with p100 cartridges can be worn. In the event, it becomes difficult to breathe comfortably or if the filter gets damaged, pause work, and change out the filter cartridge. In some cases, it may be necessary to change the type of respiratory protection device being used to an air-supplied respirator.

9.7 TREATMENT OF RESULTANT ILLNESS

First aid measures should be taken while waiting to transfer the worker to hospital. In case of accidental inhalation of a chemical mutagen by a worker, move the worker to an area with fresh air, and get emergency medical attention. If swallowed, get medical attention immediately. For skin exposure, remove contaminated clothing from the worker, and wash skin with plenty of soap and water for at least 15 minutes; get medical attention. For eye exposure, wash eyes with plenty of water for at least 15 minutes, lifting upper and lower eyelids occasionally, and then get medical attention.

REFERENCES

1. Babushok, D. V., & Kazazian Jr, H. H. (2007). Progress in understanding the biology of the human mutagen LINE-1. *Human Mutation, 28*(6), 527–539.
2. Salem, M. S. Z. (2016). Pathogenetics. An introductory review. Egyptian Journal of Medical *Human Genetics, 17*(1), 1–23.

Chapter 10

Chemical agents and occupational-related diseases

10.1 INTRODUCTION

Chemicals are used extensively and are an integral part of industry operations and our daily lives. The chemical products discussed throughout this book have uses that are beneficial and capable of enhancing the quality of life. However, these same chemicals can produce life-changing illnesses and even death. A substantial cadre of products are produced from chemicals including detergents, plastics, paints, and pharmaceutical, and the list goes on. Although some of these chemicals appear to be harmless because of their beneficial use, they can result in injuries and illness upon coming in contact with humans. On the other hand, some health effects are not immediate and only may occur after prolonged exposure. Raw materials from many sources are converted by the chemical industry into substances that are used in the production of products such as cosmetics, detergents, soaps, drugs, dye pigments, explosives, fertilizers, and a host of other useful products. These resultant products or the production process can create exposure hazards for workers and users.[6]

Chemical substances that have the capability to result in a physical or health hazard to humans are considered hazardous. Chemical substances due to their properties may be toxic, explosive, flammable, reactive, oxidizing, or corrosive. Exposure to these substances by different routes of entry, including inhalation, dermal absorption, or ingestion, can lead to adverse health effects.[5]

10.2 OCCUPATIONAL DISEASES FROM CHEMICAL EXPOSURES

Occupational diseases (ODs) can and often inflict considerable costs to workers and their family and reduce worker productivity. ODs and injuries are credited for the loss of approximately 4% of global GDP annually (based on direct and indirect costs).[2] OD refers to any disease contracted as a result of exposure to primarily chemicals arising from the work environment and

DOI: 10.1201/9781003220114-10

Table 10.1 Chemical pathways

Pathways	Health impact	Examples
Irritants	Produces reversible inflammatory changes of the skin, eyes, or mucous membranes of the respiratory tract. Can be corrosive.	Isopropyl alcohol, acetone
Sensitizers	Capable of producing an allergic response. With repeated exposure, the immune system can be sensitized and subsequently evoke an allergic response to the allergen.	Respiratory tract sensitizers: isocyanates, natural rubber latex proteins, or ethylene diamine can cause asthma. Skin sensitizers: nickel, fragrances, chromates in cement, formaldehyde, and glutaraldehyde can cause allergic contact dermatitis.
Genotoxicity	Can damage and alter genetic materials within the cells, which may cause carcinogenicity or birth defects.	
Carcinogens	Capable of causing cancers in humans.	
Mutagens	Can cause changes in the DNA of cells (mutations), which may result in various diseases or abnormalities in future generations.	Chloroform and ethylene oxide, benzene, lead, and vinyl chloride
Teratogens	Chemicals that cause birth defects, abnormalities, developmental delays, or fetal death but cause no damage to the mother, sterility, reduced fertility, and spontaneous abortion.	Methyl mercury, lead, and xylene
Reproductive toxins	Chemicals that can affect the ability of men and women to produce offspring.	

Source: https://onlinelibrary.wiley.com/doi/epdf/10.1002/9781118410868.

activities. Diagnosis of OD requires establishing of the causal relationship between chemical exposure in the work environment or work activity and a specific disease.[1]

There are many ways in which a chemical can cause harm or disease in humans. These pathways of potential health impact and examples of chemicals that can lead to specific health conditions are shown in Table 10.1.

10.3 PREVENTION OF OCCUPATIONAL-RELATED DISEASES RESULTING FROM CHEMICALS

In the work environment, the ODs that are the result of specific chemical agents are mostly preventable. Prevention can be addressed through three levels (Table 10.2). These levels include primary prevention that prevents

Table 10.2 Prevention of chemical related diseases

Primary	Secondary	Tertiary
Hierarchy of controls	Screening: medical surveillance	Medical treatment rehabilitation compensation/ medical disability
Elimination • Substitution • Engineering controls • Administrative controls • personal protective equipment (PPE)		
Monitoring • Personal • Area • Biological • Environmental		

Source: https://onlinelibrary.wiley.com/doi/epdf/10.1002/9781118410868.

the occurrence of a disease or injury through the elimination of the causal agent preventing it from causing damage to the worker; secondary prevention detects injury or illnesses in its early stage to stop the progression of the disease before symptoms are visible, and tertiary prevention is applied to workers where the disease is present and the individual requires treatment or some type of rehabilitation to minimize disability and improve the quality of life.

Once an OD resulting from chemical exposures has been diagnosed, in addition to management of the injury or illness, additional measures should be approached. These measurements should include the following:

- Remove immediately the exposed worker from further exposure.
- Remove other workers from the environment until it is deemed safe for reentry.
- Identify other workers who have been exposed.
- Investigate and control the source of exposure.
- Notify relevant authorities.
- Educate workers on the nature of the exposure and what is being done to prevent reoccurrence.
- Consult with occupational medical doctor (OMD).

Prevention of OD begins with having the knowledge of the chemicals that are used in the respective work environment and their potential capability to produce health effects. The management and prevention of ODs depend on the reduction of hazardous exposures in the workplace and educating workers, managers, and physicians on hazards. ODs are often incorrectly diagnosed because they mimic diseases from other causes and because many physicians are not trained in occupational medicine. Therefore,

Table 10.3 Occupational diseases caused by exposure to chemicals in the work environment[3]

Beryllium	Cadmium	Phosphorus	Chromium
Manganese	Arsenic	Mercury	Lead
Fluorine	Carbon disulfide	Halogen derivatives of aliphatic or aromatic hydrocarbons	Benzene
Carbon disulfide	Nitro- and amino-derivatives of benzene or its homologues	Nitroglycerine or other nitric acid esters	Alcohols, glycols, or ketones
Asphyxiants like carbon monoxide, hydrogen sulfide, hydrogen cyanide	Acrylonitrile	Oxides of nitrogen	Vanadium
Antimony	Hexane	Mineral acids	Pharmaceutical agents
Nickel	Thallium	Selenium	Copper
Platinum	Tin	Zinc	Benzoquinone
Ammonium	Isocyanates	Pesticides	Sulfur
Organic solvents	Latex and latex-containing products	Chlorine	

opportunities for prevention and treatment may be lost[4]. Table 10.3 lists some of these chemicals that can be encountered in the workplace and cause life-threatening diseases.

Chemicals found in the workplace that can cause occupational-related diseases can be placed in three categories based on the effects. The effects include occupational cancers and skin and respiratory diseases.

10.3.1 Occupational cancers

The period between the initial exposure to the substances that can cause cancer and development of cancer can be more than 10–15 years. It can potentially be longer as in the case of asbestos-related mesothelioma, which can take 40–50 years to develop. Susceptibility to occupational carcinogens is higher when exposure occurs at a younger age, or if there are combined exposures such as smoking and asbestos[1]. Carcinogens are covered in more detail in Chapter 8. This section represents additional information that is supplementary to the information contained in this chapter.

A list of some known and suspected carcinogenic substances is given in Table 10.4. Many chemicals are classified as being suspected of having the capability to cause cancer. The classification exists in these cases because

Table 10.4 Confirmed and suspected occupational carcinogen

Target organ	Confirmed	Suspected
Bone	–	Beryllium
Brain	Vinyl chloride	–
Gastroenteric track	Asbestos	–
Hematopoietic tissues (leukemia)	Styrene butadiene and other rubber manufacture substance	–
Kidney	Coke oven emissions	Lead
Larynx	Asbestos, chromium	–
Liver	Vinyl chloride	Aldrin, carbon tetrachloride, chloroform, DDT, dieldrin, heptachlor, PCBs, trichloroethylene
Lung	Arsenic Asbestos Bis (chloromethyl) ether Chloromethyl methyl ether Chromates Coke oven emissions Mustard gas Nickel Soot and tars Uranium Vinyl chloride	Beryllium Cadmium Chloroprene Lead
Lymphatic tissue	–	Arsenic Benzene
Nasal cavity	Chromium, isopropyl oil, nickel, wood dust	–
Pancreas	–	Benzidine, PCB's
Pleural cavity	Asbestos	–
Prostate	–	Cadmium
Scrotum	Soot and tars	–
Skin	Arsenic, coke oven emissions, cutting oils, soot and tars	Chloroprene
Urinary bladder	4-minobipheynl Benzidine B-Naphthylamine	Auramine 4-Nitrodiphenyl Magenta

not enough data are available to definitively classify as a known carcinogen. Therefore, when working with or in environments where chemicals are present, appropriate controls should be put in place to prevent exposures.

10.3.2 Occupational skin diseases

Occupational skin diseases can be caused by chemical agents, mechanical trauma, physical agents, as well as biological agents. However, chemicals acting as an irritant or a sensitizer are the most common cause of occupational skin disease. Contact dermatitis is recognized as the most common type of occupational skin disease and is characterized by an inflammation of the skin due to exposure to specific chemicals. The two main types of contact dermatitis are irritant contact dermatitis (ICD) and allergic contact dermatitis (ACD). ICD is a non-immunologic reaction of skin inflammation to an irritant upon exposure. The reaction is typically localized to the site of contact where the damage occurred. It may be caused by acute exposure to extremely irritating agents such as acids, bases, oxidizing/reducing agents, or chronic cumulative exposure to mild and weak irritants such as water, detergents, or weak cleaning agents. ACD is a delayed hypersensitivity reaction activated by dermal contact with a skin sensitizer (allergen) in which the individual must be first sensitized to the allergen. Subsequent exposure of allergens to the skin induced an immuno-logic reaction resulting in inflammation of the skin. The reaction is not limited to the site of contact and can result in systemic responses and skin rashes at other locations on the body. Common allergens that can be found in the workplace that has the potential to cause ACD include industrial compounds such as metals, epoxy, and acrylic resins, agrochemicals such as pesticides and fertilizers, and latex.[1]

ODs of the skin are common primary for two reasons:

- The skin has a large surface area that is available for contact exposure.
- The work environment contains enormous natural and artificial materials that are capable of exerting insults to the skin.

For these reasons, the list of causal agents is diverse; however, chemicals are by far the most frequent cause of skin disorders.

Organic and inorganic chemicals are all around us in industry, on the farm, and in our homes. These chemicals act as primary irritants, allergic sensitizers, or photosensitizers and often lead to eczematous dermatitis. Anyone who works is a potential candidate for developing occupational dermatitis from the chemicals used in the work environment. Although the potential exists, not all workers are affected by dermatological disorders. There are several intrinsic factors that influence the behavior of the skin when exposed to materials; the most important is the individual's own natural defense. The outer layer of the skin is the thickest and is found on the palms and soles of the feet, and it is a protein known as keratin. The keratin layer of skin is the most important line of cutaneous defense. In a limited capacity, it offers some resistance to mild acids and water.[6] Table 10.5 lists occupations and some agents frequently associated with producing dermatitis.

Table 10.5 Occupations and dermatitis producing agents[6]

Abrasive wheel maker
Carborundum, emery
resin glue

Aircraft workers
Adhesives (resins), alkalis,
  bichromates, chromates,
  chromic acid, cutting fluids,
  cyanides, epoxy resins,
  hydraulic fluids, hydrofluoric
  acid, lubricants, nitric acid,
  oils, paints, solvents

Animal handlers
Cleaners & detergents,
  insecticides, germicides,
  pesticides,

Painters
Acrylics, epoxies, paint
  removers, pigments,
  solvents

**Automobile workers
(Assembly)**
Adhesives, asbestos,
  antifreeze, brake fluids,
  gasoline, hydraulic fluids, oils,
  solvents

**Automobile worker
(body)**
Adhesives abrasives,
  alkalis, lead paints,
  solvents

**Automobile workers
(mechanic)**
Acids, adhesives, alkalis,
  antifreeze, brake fluids,
  epoxy resins, gasoline,
  hydraulic fluids, lubricants,
  solvents, thinners

Battery makers
Alkali, cobalt, epoxy sealer,
  mercury, nickel, solvents,
  sulfuric acid, zinc chloride

Bleachers
Borax, chlorine
  compounds,
  hydrochloric acid,
  hydrogen peroxide,
  oxalic acid, potassium
  hydroxide, sodium
  hydroxide, solvents

Brick masons
Cement, chromates, epoxy
  resins, lime,

Bronzers
Acetone, ammonia,
  ammonium sulfide, amyl
  acetate, antimony sulfide,
  arsenic, arsine, benzine,
  benzol, cyanides,
  hydrochloric acid, lacquers,
  mercury, methyl alcohol,
  petroleum hydrocarbons,
  phosphorus, resins, sodium
  hydroxide, turpentine,
  varnishes,

Cabinet makers
Bleaches, glues, resin and
  casein, oils, polishes,
  rosin, shellac, solvents,
  stains

Cement worker
Cement, chromates, cobalt,
  epoxy resins, lime, resins

Chrome platers
Chromium compounds,
  solvents, metal cleaners,
  sulfuric acid

Coal tar workers
Anthracene oil, benzol,
  coal tar, creosote,
  cresol, naphtha, solvents

Construction worker
Adhesives, resin, cement,
  concrete, creosote,
  gasoline, glass fiber, oils,
  paints, pitch, solvents,
  solvents

Cotton sizers
Acids, aluminum salts, arsenic
  salts, calcium, salts,
  dicyandiamide formaldehyde

Degreasers
Alkalis, chlorinated
  hydrocarbons solvents,
  petroleum solvents

(Continued)

Table 10.5 (Continued) Occupations and dermatitis producing agents[6]

Dentists
Mercury & metallic amalgams, oil of clove, resins, waxes

Disinfectant makers
Carbolic acid, chloride of lime, chlorinated phenols, chloride, cresol, formaldehyde, iodine, mercurial, quaternary ammonium compounds, surfactants. Zinc chloride

Dry cleaners
Acetic acid, ammonia, amyl acetate, benzine, carbon tetrachloride, methanol, nitrobenzene, perchloroethylene, sizing chemicals, Stoddard solvents, trichloroethylene, turpentine

Dye makers
Antimony compounds, benzine, calcium salts, coal tar products, cresol, dextrins, dye intermediates, ferrocyanides, formaldehyde, hydroquinone, lead salts, potassium chlorate

Dyers
Acids, alkalis, bleaches, detergents, synthetic dyes, mercurial salts, solvents, zinc chloride

Electric apparatus makers
Acids, asbestos chlorinated diphenyl, chlorinated naphthalene enamels, epoxy resins, solvents, synthetic waxes

Electroplaters
Acids, alkalis, benzene, chromic acid, lime, nickel, potassium cyanide, zinc chloride, zinc cyanide, chlorinated waxes

Embalmers
Formaldehyde, mercury, oil of cinnamon, oil of clove, phenol, thymol, zinc chloride

Engravers
Acids, alkalis, chromic acid, ferric chloride, potassium chloride, solvents

Etchers
Acids, alkalis

Explosive worker
Ammonium salts, mercury compounds, nitroglycerin, picric acid, pentaerythritol tetranitrate, trinitrotoluene

Farmers
Fertilizers, disinfectants, fungicides, lubricants, oils, paints, pesticides, solvents, wood preservatives

Felt hat makers
Acids, dyes, hydrogen peroxide, mercuric nitrate, sodium carbonate

Fertilizer makers
Acids, ammonium compounds, calcium cyanamide, fluorides, lime, nitrates, pesticides, phosphates, potassium salts

Florists
Fertilizers, herbicides

Foundry worker
Acids, lime, resin binder, solvents

Fur processors
Acids, alkalis, bleaches, chromates, dyes, formaldehyde, lime, oils, salt

Furniture polishers
Acids, alkalis, benzene, methyl alcohol, naphtha, pyridine, rosin, solvents, stains, turpentine, waxes

Galvanizers
Acids, alkalis, zinc chloride

Gardeners
Fertilizers, fungicides, herbicides, insecticides

Glass workers
Arsenic, borax, boric acid, hydrofluoric acid, lead compounds, lime, metallic oxides, petroleum oils, resins

(Continued)

Table 10.5 (Continued) Occupations and dermatitis producing agents[6]

Histology technicians
Alcohol, aniline, benzol, epoxy resins, formaldehyde, mercury bichloride, osmium dichromate, stains, toluene, xylene, waxes

Ink makers
Anti-skinning agents, chrome pigments, cobalt compounds, detergents, dyes, mercurial pigments, resins, solvents, turpentine

Janitors
Detergents, disinfectants, polishes, solvents, waxes

Jewelers
Acids, adhesives, chromium, cyanides, mercury, mercury solvents, nickel, solder flux, rouge

Chemical laboratory worker
Acids, alkalis, chromates, organic chemicals, solvents

Laundry workers
Alkalis, bleaches, fungicides, chemical dust, fiber glass, optical brightener, soaps

Machinists
Aqueous cutting fluids, chlorinated cutting oils, chromates, germicides, lubricants, rust inhibitors solvents, cutting fluids

Metal polishers
Acids, alkalis, ammonia, naphtha, pine oil, potassium cyanide, soaps, soluble oils, solvents, triethanolamine, waxes

Nickel platers
Acids, alkalis, degreasers, detergents, nickel sulfate, zinc chloride

Nitroglycerin makers
Ethylene glycol dinitrate, nitric acid, nitroglycerin, sodium carbonate, sulfuric acid

Nurses
Anesthetics, detergent, disinfectants, drugs, ethylene oxide, tranquilizers

Oil field workers
Acids, alkalis, brine, crude petroleum, explosives, lubricating oils

Paraffin workers
Paraffin, paraffin distillates, solvents,

Pencil makers
Aniline dyes, chromium pigments lacquer, resins, solvents, pyridine, methyl violet, glues

Petroleum refinery workers
Acids, alkalis, aluminum chloride, gas oil, gasoline, hydrofluoric acid, kerosene, paraffin, petroleum, tar, waxes

Photographers
Acids, alkalis, chromates, hydroquinone, methyl para-aminophenol sulfate, para-aminophenol, paraformaldehyde, paraphenylenediamines, pyrogallic acid, sodium hypochlorite, sodium sulfide, turpentine

Physicians
Adhesives, anesthetics, drugs, detergent, tranquilizers

Plumbers
Adhesives, caulking compound, fluxes hydrochloric acid Solvents, tar, zinc chloride

Refrigeration workers
Ammonia, chromates, brine, dry ice, ethyl bromide, ethyl chloride, methyl chloride, sulfur dioxide

Rayon workers
Acetic anhydride, acids, alkalis, ammonium sulfide, bleaches, calcium bisulfite, carbo disulfide, coning oils, sodium cyanide, sodium sulfide, solvents

Road workers
Cement, asphalt, concrete, epoxy resin, herbicides, paint, parasites, tar

(Continued)

Table 10.5 (Continued) Occupations and dermatitis producing agents[6]

Rocket fuel handler	Shipyard workers	Shoemakers
Aniline, boron hydrides, chlorine trifluoride, dimethylhydrazine, ethyl oxide, fuming nitric acid, gasoline, hydrazine, hydrogen fluoride, hydrogen peroxide, kerosene, liquid oxygen	Chlorinated diphenols, chlorinated naphthalenes, chromates, paint removers, paint thinners, paints, resins, solvents, tar	Adhesives, ammonia, amyl acetate, amyl alcohol, aniline dyes, benzene, benzol, hexane, naphtha, resins, waxes
Soldiers	**Stone workers**	**Tannery workers**
Acids, cyanides, fluxes, hydrazine salts, rosin, zinc chloride	Cement, lime	Acetic acid, alum, ammonium chloride, arsenic salts, benzol, brine, calcium hydrosulfide, chromium compounds, dimethylamine, mineral dyes, formaldehyde, lime, oils, sodium hydroxide, sodium sulfide, solvents, sulfuric acid
Taxidermists	**Upholsterers**	**Watch makers**
Arsenic salts, calcined alum, mercuric chloride, solvents, zinc chloride	Glues, lacquer, lacquer solvents, methyl alcohol	Acids, chromates, metal polishes, nickel, potassium cyanide, solvents
Wood preservers	**Wood workers**	
Chlorophenols, chromates, copper compounds, creosote, cresol, mercuric chloride, resins, tar, phenylmercuric compounds, zinc chloride, zinc sulfate	Acid bleaches, epoxy glues, amino resin glues, formaldehyde, lacquers, oil stains, solvents, varnishes	

10.3.3 Occupational respiratory disease

The annual number of deaths attributed to occupational respiratory diseases has been estimated to be around 318,000 for chronic obstructive pulmonary diseases (COPDs), 38,000 for asthma, and 30,000 for pneumoconioses. A large percentage of adult asthma is attributed to occupational exposure to specific agents. Occupational asthma is characterized by variable airflow limitations and airway hyperresponsiveness resulting from specific agents inhaled. Immunologic asthma develops after a latent period of exposure to sensitizers such as isocyanates, wood dust, soldering, and welding fumes. Non-immunologic asthma, commonly known as

Table 10.6 Some acute and chronic occupational respiratory diseases[1]

Acute occupational respiratory diseases	Chronic occupational respiratory diseases
Occupational asthma	Silicosis
Acute respiratory reactions to irritant gases	Coal worker's pneumoconiosis
	Asbestos and other asbestos-related diseases
Acute systemic reactions to metal fumes, polymer fumes, and organic dust	Metal lung disease
	Beryllium disease
Hypersensitivity pneumonitis	Chronic obstructive lung disease

reactive airway disease, can develop rapidly without a latency period. This form of asthma is usually associated with exposure to high concentrations of chemical irritants. Accidental inhalation of irritant gas such as chlorine, fumes, and vapors can lead to reactive airway disease. Pneumoconiosis is characterized with diffused fibrotic reaction of the lung tissues that occurs with prolonged inhalation of mineral dust. The most common type of pneumoconiosis silicosis is caused by inhalation of dust containing crystalline silicon dioxide. Table 10.6 lists the common acute and chronic occupational respiratory diseases that result from exposure to toxic chemical agents.[1]

Many chemicals encountered in the work environment can result in pneumoconiosis. The pneumoconioses are a group of lung diseases caused by inhalation of various dusts. The main cause of pneumoconioses is workplace exposure[7] Table 10.7 contains some agents along with the type of respiratory impairment seen in workers that have been exposed to the various chemical agent.

Respiratory diseases are not only developed from work being performed in the work environment, but it can also arise from breathing chemically polluted air in the external work environment. Air pollution is a well-known source that negatively affects human health and causes chronic inflammation that can lead to the development and exacerbation of respiratory diseases. Various human health issues have been raised due to urbanization, industrialization, population growth, climate change, and aging of the working population. According to the World Health Organization, air pollution accounts for greater than 6.5 million deaths worldwide per year. Air pollution, particularly particulate matter 2.5 (PM2.5), is a leading cause of several adverse health effects, including pulmonary diseases, and leads to 3.3 million premature deaths per annum worldwide. Both air pollution and occupational air pollution can be avoided, and primary prevention is critical for their control and prevention. Exposure to respirable dust, fumes, vapors, gases, volatile organic compounds, chemicals, and metals is hazardous and can cause occupational lung diseases. Occupational lung diseases include asthma, COPD, pneumoconiosis, hypersensitivity pneumonitis, and malignant diseases (lung cancer and malignant mesothelioma), with work-related asthma (WRA) being the most common.[8]

Table 10.7 Chemicals and associated Pneumoconioses[6]

Agent	Type of pathology	Type of respiratory impairment
Silica		
Simple	Nodular fibrosis	Restrictive, diffusion
Complicated	Conglomerate nodular fibrosis Nodular fibrosis	Restrictive, obstructive, diffusion Restrictive, diffusion
Hematite		
mixed dust	Nodular fibrosis	Restrictive, diffusion
Iron and silica	(rarely conglomerate nodular fibrosis)	
Silicates talc	Nodular fibrosis (rarely conglomerate nodular fibrosis)	Restrictive, diffusion
Kaolin		
Bentonite		
Diatomie		
Tripoli		
Fuller's earth		
Mica		
Sillimanite	Nonspecific bronchitis	Obstructive
Cement		
Coal	Peribronchiolar macules, focal	Obstructive (small airways)
Simple	emphysema	
	Conglomerate nodular fibrosis	Obstructive, restrictive,
Complicated	Peribronchiolar macules, focal emphysema	diffusion Obstructive (small airways)
Graphite	Interstitial fibrosis	Restrictive, diffusion
	Interstitial fibrosis	Restrictive, diffusion
Aluminum	Interstitial fibrosis	Restrictive, diffusion
Asbestos	Interstitial fibrosis	
Beryllium	(granulomata) Interstitial fibrosis	Restrictive, diffusion
Tungsten carbide	Interstitial fibrosis	Restrictive, diffusion

REFERENCES

1. Lim, J. W., & Koh, D. (2014). Chemical agents that cause occupational diseases. The Wiley Blackwell Encyclopedia of Health, Illness, Behavior, and Society. https://onlinelibrary.wiley.com/doi/epdf/10.1002/9781118410868.wbehibs399.

2. Davoodi, S., Haghighi, K.S., Kalhori, S., Hosseini, N.S., Mohammadzadeh, Z., & Safdari, R. (2017). Occupational disease registries – characteristics and experiences. Acta Informatica Medica, 25(2), 136–140. https://doi.org/10.5455/aim.2017.25.136-140.

3. International Labor Organization (2010). List of Occupational Diseases. International Labor Organization. http://www.ilo.org/wcmsp5/groups/public/@ed_protect/@protrav/@safework/documents/publication/wcms_150323.pdf.

4. Landrigan, P.J., & Baker, D.B. (1991). The recognition and control of occupational disease. JAMA, 266(5),676–80.

5. National Institute of Occupational Safety and Health (2018). Emergency Response Resources. National Institute of Occupational Safety and Health, U.S. Department of Health and Human Services. https://www.cdc.gov/niosh/emres/default.html.

6. Key, M. M., & Ede, L. (1977). Occupational diseases: a guide to their recognition. US Department of Health, Education, and Welfare, Public Health Service, Center for Disease Control, National Institute for Occupational Safety and Health. https://books.google.com/books?hl=en&lr=&id=oNe8sjw3agcC&oi=fnd&pg=PA3&dq=+Occupational+diseases,+A+guide+to+their+recognition.&ots=b5v3L5_6Sk&sig=NzwAdW5t_Osv1eXrdLvfLKQEDhc#v=onepage&q=Occupational%20diseases%2C%20A%20guide%20to%20their%20recognition.&f=false

7. National Institute of Occupational Safety and Health (2011). Workplace Safety and Health Topics: Pneumoconioses. National Institute of Occupational Safety and Health, U.S. Department of Health and Human Services. https://www.cdc.gov/niosh/topics/pneumoconioses/default.html.

8. Nishida, C., & Yatera, K. (2022). The Impact of ambient environmental and occupational pollution on respiratory diseases. International Journal of Environmental Research and Public Health, 19(5), 2788. https://doi.org/10.3390/ijerph1905278.

Appendix A: Exposure and risk assessment strategy

Exposure assessment is the process used to evaluate and determine the potential risk of exposure to a particular substance or product and the potential harm that can be inflected of the exposed individual or population. Exposure assessment is an important step in the process to identify, quantify, and mitigate exposure risks.

This appendix provides some tools to assist in the completion of an exposure assessment strategy and processes, identify, and mitigate risk to reduce or eliminate harmful exposure.

Each employer is required by law to protect workers from harmful hazardous conditions in the workplace. Exposure assessment and risk evaluation are viable processes that have proven to be beneficial in protecting workers from workplace hazards.

Exposure Pathway Evaluation Worksheet

Identify task or process _____

_____

Identify workgroup _____

Skin contact:	Yes	No
1. Will skin be completely covered with the appropriate PPE?	yes	no
2. Is there a potential for skin contact?	yes	no

Inhalation:	Yes	No
1. Are chemicals used in a closed system? (if yes, no inhalation hazard)	yes	no
2. Is the chemical in the form of a dusts, mists, fumes, vapors, and gas	yes	no
3. If yes to #1, can the form be dispersed in the environment and become airborne?	yes	no
4. Does the chemical have a high vapor pressure?	yes	no

Ingestion:		
1. Are food products consumed in the area?	yes	no
2. Is the work area located near break area and a potential exist for contamination?	yes	no

Injection:		
1. Is there a potential for a puncture hazard?	yes	no

Note: answering yes to any of the above question represents a potential exposure pathway unless otherwise noted

---

Exposure Assessment Worksheet (example)

Location: _____   Date: _____

Task/Process description: _____
_____
_____

Similarly Exposed Group (s): _____

Potential Hazard	OEL	Exposure Route	Exposure Frequency	Exposure Duration	Controls	PPE	Exposure rating	Air sampling required (Y/N)

Exposure rating scheme:

1- Continuous
2- Intermittent (2 – 3 times per day)
3- Infrequently (1 – 2 times per year)
4- Occasionally (1 – 2 times every 5 years)

Comments:

Recommendations:

Future monitoring and evaluation: -
_____
_____

Note: Attached a copy of the monitoring strategy

Investigator/Assessor:

Title: _____

Investigator/Assessor signature: _____

Inhalation Exposure Pathway Evaluation Worksheet

Date:

Document Number:

Identify task or process: (include time needed to complete task, frequency of task performance)

_____

_____

Identify the chemical (s) of concern: _____

_____

Identify workgroup: _____

Has engineering controls been implemented? Yes or No   If implemented is it effective in removing vapors and particulate for the environment Yes or No

What measures can be implemented to remove inhalation potential? _____

Will PPE be required? If so, what type? Specify

Job/Task		JSA#	Start Date:	End Date:
Department/Work Group:	Building:	Area:		Other:
Work Scope: (attach diagrams and permits)				

Task #	Activity	Hazard	Control

Worker Acknowledgement: a commitment to perform work as documented utilizing all safety precautions. A stop work will be initiated if work cannot be performed as document.

Print Name	Signature	Date
Print Name	Signature	Date
Print Name	Signature	Date
Print Name	Signature	Date
Print Name	Signature	Date
Print Name	Signature	Date

Approval: (indicating that the JSA has been reviewed and in agreement with the entire content and the work to be performed)

Work Group Supervisor:	Signature:	Date:
Environment Safety & Health:		
Area Manager:		
Other:		

PPE Selection Form

Performed by: _____   Date: _____

Process/Task: _____   Work group: _____

Source	Hazard	PPE Required	Special instructions
Chemicals ☐ Biological Agent ☐ Physical process ☐	Eye or Face injury potential Impact from flying objects ☐ Chemical splash in eyes ☐ Facial skin chemical contact ☐ Skin contact potential Hand ☐ Body - whole ☐ Extremities ☐ Face ☐	Safety glasses ☐ Safety glasses with side shields ☐ Safety goggles ☐ Face shield ☐ Latex gloves ☐ Rubber gloves ☐ Chemical resistant gloves ☐ Chemical apron ☐ Tyvek ☐ Lab coat ☐	
Task generating airborne ☐ fibers, dust, fumes, mist, or vapor	Inhalation exposure above OEL ☐ Inhalation exposure below OEL	Respiratory protection Organic vapor cartridge ☐ Particulate cartridge ☐ Other ☐	Request exposure assessment to be completed of the airborne constituents
High noise from equipment or task performance ☐	Inhalation exposure above OEL ☐ Inhalation exposure below OEL	Ear plugs ☐ Earmuffs ☐ Other ☐	Noise survey to verify level and PPE requirement
Non ionizing radiation Lasers ☐ Welding ☐	Potential burns: Eyes ☐ Skin ☐	Laser safety glasses ☐ Welder's helmet ☐ Other ☐	Request a non-ionizing survey
General safety – physical hazards resulting from equipment, process, or material ☐	Foot injury potential ☐ Impact potential – head ☐ Electrical ☐	Safety shoes ☐ Hard hat ☐ Cut resistant gloves ☐ Electrical gloves ☐ Other ☐	
Other ☐	Cold or heat extremes ☐	Thermal gloves/clothing ☐	

APPENDIX A.5

Helpful Information

1. **Vapor pressure:** Vapor pressure measures chemical's volatility. Chemicals with higher vapor pressure are more likely to enter air; therefore, these products must be used in a manner that their emission to air is minimized. High vapor pressures and high evaporation rates also lead to lower flash points and higher flammability (https://www.chemsafetypro.com/Topics/CRA/Vapor_Pressure.html).
2. Effort to lessen the potential of hazardous dermal (skin) exposure.
   - Substitute a less hazardous chemical.
   - Consider re-designing the work process to avoid splashes and skin contact. Where that is not feasible, personal protection in the form of chemical protective gloves, apron, or other protective clothing should be selected.
   - Invoke good housekeeping to avoid the accumulation of stable, low volatility, dermally toxic contaminants on horizontal surfaces.
   - Review published breakthrough information from glove manufacturers and lab test data when selecting appropriate hand protection (https://www.osha.gov/dermal-exposure/control-prevention)

3. Examples of administrative controls
   a. Training workers on hazard identification and control measures
   b. Training workers on safe procedures for working around the hazard
   c. Rearranging steps in a job task to prevent the worker from coming into contact with hazards
   d. Developing standardized safe work practices
   e. Requiring workers in hot environments to take breaks in a cool environment at some periodicity
   f. Proper housekeeping to eliminate trip and fall hazards
   g. Posting visual reminders for workers to inform them which areas are prohibited from entering
4. Examples of engineering controls
   a. Using wet methods to control dust when drilling or grinding
   b. Installing enclosures to keep workers from coming in contact with hazardous chemicals or rotating equipment parts
   c. Use glove boxes to enclose and isolate operations
   d. Interlocking systems to prevent contact during operation
   e. Use ventilation systems such as fume hoods, etc.

Appendix B: Exposure monitoring

Employers are responsible for developing and implementing a comprehensive occupational safety and health program to prevent occupational injuries, illnesses, and deaths. Safety and health programs should be developed and implemented as part of an employer's management system, with strong management commitment, employee involvement, and occupational safety and health expertise to be successful.[3]

Employers are charged with ensuring that workers are protected while working in a work environment as such they are also charged with developing and implementing a comprehensive occupational health and safety program. This program should be centered around prevention of injuries, illness, and death in the workplace. A good monitoring strategy is a necessary part of a comprehensive occupational health and safety program. This section provides tools that can be used to develop and implement an exposure monitoring program that can serve as a foundation for a comprehensive exposure monitoring program (Table B.1).

A workplace exposure monitoring program should have clear, stated goals. Exposure assessment strategies must be site-specific and should be developed to accomplish the following goals:

- To determine employee exposure to chemicals used in the workplace.
- To evaluate the effectiveness of work practices and engineering controls.
- To facilitate selection of appropriate personal protective equipment, if required.

The monitoring strategy should also assess the effectiveness of engineering controls, work practices, PPE, training, and other factors in used controlling exposures; identify areas or tasks that are associated with higher exposures where additional control efforts and/or sampling are needed and; determine how changes in production may impact employee exposures.[3]

Table B.1 Exposure monitoring program elements

Elements	Important factors
Objective of sampling	Objectives can include characterizing compounds present in workplace air or in bulk materials; ensuring compliance with OELs; evaluating the effectiveness of engineering controls, work practices, PPE, training, or other methods used to control exposures; identifying areas, tasks, or jobs with higher exposures that require additional exposure control; evaluate high-risk job categories to ensure that exposures do not exceed OELs; measure exposures of employees who report symptoms or illnesses.
Specific agent to sample	Determining what to sample generally requires knowledge of the process or task.
Whom and where to sample	Selecting whom or where to sample depends on the sampling objectives. When sampling to determine whether employee exposures are below the OELs, a focused or compliance sampling strategy that targets employees perceived to have the highest exposures. Area sampling is also useful for determining sources of airborne contaminants and identifying the worst-case chemical concentrations. Selection of which employees or work locations should be sampled can help to characterize potential exposures and areas of concern.
How to sample	Consulting with the analytical laboratory before sampling to ensure that the appropriate sampling protocol and media are used and measurement methods available can meet the defined sampling requirements.
When to sample	To determine potential exposures and compliance with OELs. Exposure monitoring should be conducted whenever there is a change in production processes, controls, work practices, or other conditions, which indicate a potential change in exposure conditions.
How long to sample	TWA exposures should be determined by collecting samples over a full work shift. The sample result is then compared with OELs. Work shifts that exceed 8 hours will require extended sampling duration.
How many samples to collect	The number of samples required for an accurate exposure assessment depends primarily upon the purpose of the sampling; the number of processes, work tasks, or jobs to be evaluated; the variability inherent in the measured contaminant concentrations, sampling and analytical variability; and other factors.
Sample handling storage, and shipment	Before collecting samples, close collaboration with the analytical laboratory on handling, storage, and shipping methods required for each analyte is necessary. Samples collected should be analyzed by an accredited analytical lab such as one having American Industrial Hygiene Association Accreditation.

Source: Criteria for a recommended standard: occupational exposure to diacetyl and 2,3-pentane-dione (cdc.gov).

APPENDIX B-1

Exposure Monitoring Worksheet

Sample identifier: _____   Date: _____
Sample type (area, personal, blank) _____

Collected by: _____   Collection time: _____

Person/process sampled (name):

Work Environment

Process/task or work environment description: _____
_____

Personal protective equipment used: _____

Engineering controls used: _____

Temperature: _____ Wind speed: _____ Wind direction: _____
Humidity: _____

Equipment and Calibration Data

Sampling equipment Mfg. and model	Flow rate	Units sample start time:	
	Start: _____ End: _____ Average: _____		

Analysis and Results

Chemical/substance	Concentration and unit:	Occupational exposure limit (OEL)

Analytical laboratory: _____
Analyzed by: _____

APPENDIX B.2

QUESTIONS TO HELP DEVELOP AN EFFECTIVE EXPOSURE MONITORING STRATEGY

1. What is the task of concern?
2. How many personnel were potentially exposed?
3. How long will the task take to complete?
4. What chemical hazards in the workplace and process that workers may be exposed?
5. Are data available to quantify potential exposures?
6. What chemicals require data collection to quantify exposures?
7. What sampling media will be used to collect samples?
8. Are there any special collection and storage requirements for the samples based on the type of media selected and analytical requirements?
9. How many and what type of sampling are needed (e.g., personal, area, wipe)?
10. What laboratory method will be used to analyze the sample?
11. What accreditation does the lab has that will be analyzing the samples?
12. What are the OEL for the chemicals?
13. Are engineering controls in place? If so, have they been deemed effective at preventing overexposure?
14. Have the engineering controls in place been verified effective through sample collection and analysis?
15. Will personal protective equipment be required to prevent exposures?
16. If PPE will be used, what type? Has an evaluation been performed to determine applicability and potential effectiveness?
17. Will administrative controls be employed to prevent exposures?

Notification of employees

Employers should establish policies and procedures for the timely notification of employees of their personal and environmental monitoring results and results that are representative of their work group exposure. Communication of data to employees and management can occur in person, via e-mail (depending on the results), or through formal documentation. In either case, care must be given to the tone of the communication and ensure that is written simplistically and clearly so that it can be easily understood.

In addition, employers should inform employees of the following:

- Actions taken to prevent or reduce exposures based on the result of monitoring.
- Any change in processes that is expected to impact future exposures.
- Any changes in PPE.
- Any changes in work practices.

Employee Notification of Sampling Results

Name: _____   Date: _____

Contaminant Monitored: _____

Summary of Task Monitored: _____

_____

Results: _____   OEL _____

Summary of Results: _____

_____

_____

Recommendations: _____

_____

_____

Sample Results Reviewed and Validated by: _____

Signature: _____

Figure B.1 Employee sample notification form.

HELPFUL RESOURCES AND INFORMATION

1. ACGIH, American Conference of Governmental Industrial Hygienists. (2001). *Air-sampling instruments for evaluation of atmospheric contaminants.* 9th ed., Cincinnati, OH: American Conference of Governmental Industrial Hygienists.
2. Alston, F., Millikin, E., & Piispanen, W. (2018). *Industrial hygiene: Improving worker health through an operational risk approach.* CRC Press, Taylor & Francis Group.

3. Criteria for a recommended standard: occupational exposure to diacetyl and 2,3-pentanedione (cdc.gov).
4. FEMA. (2012). *Respiratory health and safety in the flavor manufacturing workplace.* In: *The flavor and extract manufacturers association of the United States.* Washington, DC, p. 28.
5. Gross, E.R., & Pechter, E. (2002). Evaluation. In: Plog, B.A., Ed. *Fundamentals of industrial hygiene.* 5th Ed. National Safety Council, pp. 487–519.
6. Hewett, P., Logan, P., Mulhausen, J., Ramachandran, G., & Banerjee, S. (2006). Rating exposure control using Bayesian decision analysis. *Journal of Occupational and Environmental Hygiene, 3*(10):568–581.
7. IOSH. (1977). *Occupational exposure sampling strategy manual.* Cincinnati, OH: U.S. Department of Health, Education, and Welfare, Center for Disease Control, National Institute for Occupational Safety and Health, DHEW (NIOSH) Publication No. 77–173.
8. Jennison, E.A., Kullman, G.J., & Parker, J.E. (1996). Site visits. In: Harber P., Schenker, M., & Balmes, J., Eds. *Occupational and environmental respiratory disease.* New York: Mosby, pp. 140–147.
9. Milz, S.A., Conrad, R.G., & Soule, R.D. (2003). *Principles of evaluating worker exposure.* In: DiNardi, S.R., Ed. *The occupational environment: Its evaluation, control, and management.* Fairfax, VA: The American Industrial Hygiene Association, pp. 116–128.
10. Mulhausen, J.R., & Damiano, J. (1998). *A strategy for assessing and managing occupational exposures.* 2nd Ed. Fairfax, VA: American Industrial Hygiene Association, pp. 15–56.
11. NIOSH. (1977). *Occupational exposure sampling strategy manual.* Cincinnati, OH: U.S. Department of Health, Education, and Welfare, Center for Disease Control, National Institute for Occupational Safety and Health, DHEW (NIOSH) Publication No. 77-173.
12. NIOSH. (1994). *NIOSH manual of analytical methods (NMAM).* 4th Ed. Schlecht, P.C., & O'Connor, P.F., Eds. Cincinnati, OH: U.S. Department of Health and Human Services, Centers for Disease Control and Prevention, National Institute for Occupational Safety and Health, DHHS (NIOSH) Publication 94–113 (August 1994); 1st Supplement Publication 96–135; 2nd Supplement Publication 98–119; 3rd Supplement 2003–154.
13. Snedecor, G.W., & Cochran, W.G. (1967). *Statistical methods.* Ames, IA: The Iowa State University Press, pp. 111–119.
14. Soule, R.D. (2000). *Workplace sampling and analysis.* In: Patty's industrial hygiene. 5th Ed. New York: John Wiley & Sons, pp. 265–315.

Appendix C: Personal protective equipment

C.1 PERSONAL PROTECTIVE PROGRAM

The work environment is filled with different types of hazards namely physical, biological, chemical, ergonomic, and radiological hazards. The United States Occupational Safety and Health Administration (OSHA) requires that employers protect their employees from workplace hazards that can cause injury. There is a hierarchy of hazard control that is regarded as the best practice for protecting workers from exposure to these hazards. The use of engineering or work practice controls to eliminate hazards is the first preference. When engineering, work practice, and administrative controls are not feasible or do not provide sufficient protection, employers must provide personal protective equipment (PPE) to their employees and ensure its use. PPE is equipment worn by workers to prevent or minimize exposure to hazards that cause serious workplace injuries and illnesses. These injuries and illnesses may result from contact with chemical, radiological, physical, electrical, mechanical, or other workplace hazards. PPE may include items such as gloves, safety glasses and shoes, earplugs or muffs, hard hats, respirators or coveralls, vests, and full-body suits. See Table C.1 for OSHA PPE standards.

C.1.1 Hazard assessment

The hazard assessment should begin with a walkthrough survey of the workplace to develop a list of potential hazards. When the walkthrough is complete, the employer should organize and analyze the data so that they may be efficiently used in determining the proper types of PPE required at the worksite. The employer should have knowledge of the different types of PPE available and the levels of protection offered. It is best practice to select PPE that will provide a level of protection greater than the minimum required to protect employees from hazards. The workplace should be periodically reassessed for any changes in conditions, equipment, or operating procedures that could affect occupational hazards. This periodic reassessment should also include a review of injury and illness records to spot any trends or areas of concern and take appropriate corrective action. The suitability of existing PPE, including an evaluation of its condition and age, should be included in the reassessment.

C.1.2 PPE selection

PPE should be safely designed, constructed, and maintained clean. Employers should take the fit and comfort of PPE into consideration when selecting appropriate items for their workplace. PPE that fits well and is comfortable to wear will encourage employee use of PPE. Most PPEs are available in different sizes, and selection of the proper PPE is based on employee size. Proper fitting of PPE ensures that workers are not exposed to hazards. The level of protection for the type of PPE should meet applicable requirements as developed by the relevant regulatory body such as the American National Standards Institute (ANSI). Employers should ensure that employees who provide their own PPE conform to the employer's criteria, based on the hazard assessment, OSHA requirements, and ANSI standards. OSHA requires PPE to meet the following ANSI standards:

- Eye and face protection: ANSI Z87.1-1989 (USA Standard for Occupational and Educational Eye and Face Protection).
- Head protection: ANSI Z89.1-1986.
- Foot protection: ANSI Z41.1-1991.
- For hand protection, there is no ANSI standard for gloves, but OSHA recommends that selection can be based on the tasks to be performed and the performance and construction characteristics of the glove material.
  - For protection against chemicals, glove selection must be based on the chemicals encountered, the chemical resistance, and the physical properties of the glove material.

C.1.3 Eye and face protection

Employers are required to ensure that employees have appropriate eye or face protection for hazard assessments that indicate potential exposure to the eye and face. Employers must be sure that their employees wear appropriate eye and face protection and that the selected form of protection is appropriate to the work being performed and properly fits each employee exposed to the hazard. For employees who wear prescription lenses, protective eyewear should not disturb the proper positioning of the prescription lenses so that the employee's vision will not be inhibited or limited.

Selection of eye and face PPE should include the following:

- Ability to protect against specific workplace hazards.
- Should fit properly and be reasonably comfortable to wear.

- Should provide unrestricted vision and movement.
- Should be durable and cleanable.
- Should allow unrestricted functioning of any other required PPE.

Types of eye and face PPE

Each type of protective eyewear is designed to protect against specific hazards. The following are the types of PPE:

- **Safety spectacles:** These protective eyeglasses have plastic or metal safety frames and impact-resistant lenses. Side shields are available on some models.
- **Goggles:** These are tight-fitting eye protection that completely cover the eyes, eye sockets, and the facial area immediately surrounding the eyes and provide protection from impact, dust, and splashes.
- **Welding shields:** These are constructed of vulcanized fiber or fiberglass and fitted with a filtered lens. They protect the eyes from burns caused by infrared or intense radiant light; they also protect both the eyes and face from flying sparks, metal spatter, and slag chips produced during welding, brazing, soldering, and cutting operations. OSHA requires filter lenses to have a shade number appropriate to protect against the specific hazards of the work being performed to protect against harmful light radiation.
- **Laser safety goggles:** These specialty goggles protect against intense concentrations of light produced by lasers. The type of laser safety goggles an employer chooses will depend upon the equipment and operating conditions in the workplace.
- **Face shields:** These transparent sheets of plastic extend from the eyebrows to below the chin and across the entire width of the employee's head. Some are polarized for glare protection. Face shields protect against nuisance dusts and potential splashes or sprays of hazardous liquids but will not provide adequate protection against impact hazards. Face shields used in combination with goggles or safety spectacles will provide additional protection against impact hazards.

C.1.4 Head protection

Protecting employees from potential head injuries is important for any safety program. Wearing a safety helmet or hard hat is one of the easiest ways to protect an employee's head from injury. Hard hats can protect employees from impact and penetration hazards as well as from electrical shock and burn hazards. Employers must ensure that their employees wear head protection if any of the following apply:

- Objects might fall from above and strike them on the head.
- They might bump their heads against fixed objects, such as exposed pipes or beams.
- There is a possibility of accidental head contact with electrical hazards.

Whenever there is a danger of objects falling from above, head protection must be worn. Hard hats must be worn with the bill forward to protect employees properly. In general, protective helmets or hard hats should do the following:

- Resist penetration by objects.
- Absorb the shock of a blow.
- Be water-resistant and slow burning.
- Have clear instructions explaining proper adjustment and replacement of the suspension and headband.

Hard hats must have a hard outer shell and a shock-absorbing lining that incorporates a headband and straps that suspend the shell from 1 to 1 1/4 inches (2.54 cm to 3.18 cm) away from the head. This type of design provides shock absorption during an impact and ventilation during normal wear. Protective headgear must meet ANSI Standard Z89.1–1986 or provide an equivalent level of protection.

C.1.5 Types of hard hats

There are many types of hard hats available in the marketplace. It is important for employers to understand all potential hazards when making this selection, including electrical hazards. This can be done through a comprehensive hazard analysis and an awareness of the different types of protective headgear available. Hard hats are divided into three industrial classes:

- Class A hard hats provide impact and penetration resistance along with limited voltage protection (up to 2,200 volts).
- Class B hard hats provide the highest level of protection against electrical hazards, with high-voltage shock and burn protection (up to 20,000 volts). They also provide protection from impact and penetration hazards by flying/falling objects.
- Class C hard hats provide lightweight comfort and impact protection but offer no protection from electrical hazards.

There is another class of protective headgear called a "bump hat," designed for use in areas with low head clearance. These are recommended for areas where protection is needed from head bumps and lacerations. These are not designed to protect against falling or flying objects and are not ANSI-approved. Check the type of hard hat a worker is using to make sure that the equipment provides appropriate protection. Each hat should bear a label inside the shell that lists the manufacturer, the ANSI designation, and the class of the hat.

C.1.6 Foot and leg protection

Workers who face possible foot or leg injuries from falling or rolling objects or from crushing or penetrating materials should wear protective footwear. Workers whose work involves exposure to hot substances, or corrosive or poisonous materials must also have protective gear to cover exposed body parts, including legs and feet. If an employee's feet may be exposed to electrical hazards, non-conductive footwear should be worn. On the other hand, workplace exposure to static electricity may necessitate the use of conductive footwear. Examples of situations in which an employee should wear foot and/or leg protection include:

- When heavy objects such as barrels or tools might roll onto or fall on the employee's feet.
- Working with sharp objects such as nails or spikes that could pierce the soles or uppers of ordinary shoes.
- Exposure to molten metal that might splash on feet or legs.
- Working on or around hot, wet, or slippery surfaces.
- Working when electrical hazards are present.

Safety footwear must meet ANSI minimum compression and impact performance standards in ANSI Z41-1991 or provide equivalent protection. All ANSI-approved footwear has a protective toe and offers impact and compression protection. But the type and amount of protection are not always the same. Different footwear protects in different ways. Foot and leg protection choices include the following:

- *Leggings* protect the lower legs and feet from heat hazards such as molten metal or welding sparks. Safety snaps allow leggings to be removed quickly.
- *Metatarsal guards* protect the instep area from impact and compression. Made of aluminum, steel, fiber, or plastic, these guards may be strapped to the outside of shoes.
- *Toe guards* fit over the toes of regular shoes to protect the toes from impact and compression hazards. They may be made of steel, aluminum, or plastic.
- *Combination foot and shin guards* protect the lower legs and feet and may be used in combination with toe guards when greater protection is needed.
- *Safety shoes* have impact-resistant toes and heat-resistant soles that protect the feet against hot work surfaces common in roofing, paving

and hot metal industries. The metal insoles of some safety shoes protect against puncture wounds. Safety shoes may also be designed to be electrically conductive to prevent the buildup of static electricity in areas with the potential for explosive atmospheres or non-conductive to protect employees from workplace electrical hazard

C.1.7 Hand and arm protection

Hazard assessment of the workplace may reveal that workers face potential injury to hands and arms that cannot be eliminated through engineering and work practice controls, hand and arm protection should be worn. Potential hazards include skin absorption of harmful substances, chemical or thermal burns, electrical dangers, bruises, abrasions, cuts, punctures, fractures, and amputations. Protective equipment includes gloves, finger guards, and arm coverings or elbow-length gloves.

C.1.8 Types of gloves

There are many types of gloves produced to protect from various chemicals. The user must have knowledge of the type of glove needed for the substances in which protection is being sought. Glove selection guides are available to assist with selecting the appropriate glove for the task. Selecting the wrong glove can create a sense of ease when in fact the protection needed is not being afforded. Some examples of appropriate glove selection and usage are discussed below.

C.1.9 Leather, canvas, or metal mesh gloves

- *Leather gloves* protect against sparks, moderate heat, blows, chips, and rough objects.
- *Aluminized gloves* provide reflective and insulating protection against heat and require an insert made of synthetic materials to protect against heat and cold.
- *Aramid fiber gloves* protect against heat and cold, are cut and abrasive-resistant, and wear well.
- *Synthetic gloves* of various materials offer protection against heat and cold, are cut and abrasive-resistant, and may withstand some diluted acids. These materials do not stand up against alkalis and solvents

C.1.10 Fabric and coated fabric gloves

Fabric and coated fabric gloves are made of cotton or other fabric to provide varying degrees of protection.

- *Fabric gloves* protect against dirt, slivers, chafing, and abrasions. They do not provide sufficient protection for use with rough, sharp, or heavy materials. Adding a plastic coating will strengthen some fabric gloves.
- *Coated fabric gloves* are normally made from cotton flannel with napping on one side. By coating the unsnapped side with plastic, fabric gloves are transformed into general-purpose hand protection offering slip-resistant qualities. These gloves are used for tasks ranging from handling bricks and wire to chemical laboratory containers. When selecting gloves to protect against chemical exposure hazards, always check with the manufacturer or review the manufacturer's product literature to determine the gloves' effectiveness against specific workplace chemicals and conditions.

C.1.11 Chemical- and liquid-resistant gloves

Chemical-resistant gloves are made with different kinds of rubber: natural, butyl, neoprene, nitrile, and fluorocarbon (viton); or various kinds of plastic: polyvinyl chloride (PVC), polyvinyl alcohol, and polyethylene. These materials can be blended or laminated for better performance. As a rule, the thicker the glove material, the greater the chemical resistance but thick gloves may impair grip and dexterity, having a negative impact on safety. Some examples of chemical-resistant gloves include:

- **Butyl gloves** are made of synthetic rubber and protect against a wide variety of chemicals, such as peroxide, rocket fuels, highly corrosive acids, strong bases, alcohols, aldehydes, ketones, esters, and nitro compounds. Butyl gloves also resist oxidation, ozone corrosion, and abrasion and remain flexible at low temperatures. Butyl rubber does not perform well with aliphatic and aromatic hydrocarbons and halogenated solvents.
- **Natural (latex) rubber gloves** are comfortable to wear, which makes them a popular general-purpose glove. They feature outstanding tensile strength, elasticity, and temperature resistance. In addition to resisting abrasions caused by grinding and polishing, these gloves protect employees' hands from most water solutions of acids, alkalis, salts, and ketones. Latex gloves have caused allergic reactions in some individuals and may not be appropriate for all employees. Hypoallergenic gloves, glove liners, and powderless gloves are possible alternatives for employees who are allergic to latex gloves.
- **Neoprene gloves** are made of synthetic rubber and offer good pliability, finger dexterity, high density, and tear resistance. They protect against hydraulic fluids, gasoline, alcohols, organic acids, and alkalis. They generally have chemical and wear resistance properties superior to those made of natural rubber.

- **Nitrile gloves** are made of a copolymer and provide protection from chlorinated solvents such as trichloroethylene and perchloroethylene. Although intended for jobs requiring dexterity and sensitivity, nitrile gloves stand up to heavy use even after prolonged exposure to substances that cause other gloves to deteriorate. They offer protection when working with oils, greases, acids, caustics, and alcohols but are generally not recommended for use with strong oxidizing agents, aromatic solvents, ketones, and acetates.

Several groups publish glove selection charts to aid in the selection of gloves based on the potential chemical exposure. An example of a guide is shown in the following table.

Glove selection chart

Chemical	Neoprene	Natural latex or rubber	Butyl	Nitrile latex
Acetaldehyde	VG	G	VG	G
Acetic acid	VG	VG	VG	VG
Acetone	G	VG	VG	P
Ammonium hydroxide	VG	VG	VG	VG
Amyl acetate	F	P	F	P
Aniline	G	F	F	P
Benzaldehyde	F	F	G	G
Benzene	F	F	F	P
Butyl acetate	G	F	F	P
Butyl alcohol	VG	VG	VG	VG
Carbon disulfide	F	F	F	F
Carbon tetrachloride	F	P	P	G
Chlorobenzene	F	P	F	P
Chloroform	G	P	P	P
Chloronaphthalene	F	P	F	F
Chromic acid (50%)	F	P	F	F
Citric acid (10%)	VG	VG	VG	VG
Cyclohexanol	G	F	G	VG
Dibutyl phthalate	G	P	G	G
Diesel fuel	G	P	P	VG
Diisobutyl ketone	P	F	G	P
Dimethylformamide	F	F	G	G
Dioctyl phthalate	G	P	F	VG
Dioxane	VG	G	G	G
Epoxy resins, dry	VG	VG	VG	VG
Ethyl acetate	G	F	G	F
Ethyl alcohol	VG	VG	VG	VG
Ethyl ether	VG	G	VG	G

(Continued)

(Continued) Glove selection chart

Chemical	Neoprene	Natural latex or rubber	Butyl	Nitrile latex
Ethylene dichloride	F	P	F	P
Ethylene glycol	VG	VG	VG	VG
Formaldehyde	VG	VG	VG	VG
Formic acid	VG	VG	VG	VG
Freon 11	G	P	F	G
Freon 12	G	P	F	G
Freon 21	G	P	F	G
Freon 22	G	P	F	P
Furfural	G	G	G	G
Gasoline, leaded	G	P	F	VG
Gasoline, unleaded	G	P	F	VG
Glycerine	VG	VG	VG	VG
Hexane	F	P	P	G
Hydrochloric acid	VG	G	G	G
Hydrofluoric acid (48%)	VG	G	G	G
Hydrogen peroxide (30%)	G	G	G	G
Hydroquinone	G	G	G	F
Isooctane	F	P	P	VG
Isopropyl alcohol	VG	VG	VG	VG
Kerosene	VG	F	F	VG
Ketones	G	VG	VG	P
Lacquer thinners	G	F	F	P
Lactic acid (85%)	VG	VG	VG	VG
Lauric acid (36%)	VG	F	VG	VG
Lineoleic acid	VG	P	F	G
Linseed oil	VG	P	F	VG
Maleic acid	VG	VG	VG	VG
Methyl alcohol	VG	VG	VG	VG
Methylamine	F	F	G	G
Methyl bromide	G	F	G	F
Methyl chloride	P	P	P	P
Methyl ethyl ketone	G	G	VG	P
Methyl isobutyl ketone	F	F	VG	P
Methyl methacrylate	G	G	VG	F
Monoethanolamine	VG	G	VG	VG
Morpholine	VG	VG	VG	G
Naphthalene	G	F	F	G
Naphthas, aliphatic	VG	F	F	VG
Naphthas, aromatic	G	P	P	G
Nitric acid	G	F	F	F
Nitromethane (95.5%)	F	P	F	F

(Continued)

(Continued) Glove selection chart

Chemical	Neoprene	Natural latex or rubber	Butyl	Nitrile latex
Nitropropane (95.5%)	F	P	F	F
Octyl alcohol	VG	VG	VG	VG
Oleic acid	VG	F	G	VG
Oxalic acid	VG	VG	VG	VG
Palmitic acid	VG	VG	VG	VG
Perchloric acid (60%)	VG	F	G	G
Perchloroethylene	F	P	P	G
Petroleum distillates (naphtha)	G	P	P	VG
Phenol	VG	F	G	F
Phosphoric acid	VG	G	VG	VG
Potassium hydroxide	VG	VG	VG	VG
Propyl acetate	G	F	G	F
Propyl alcohol	VG	VG	VG	VG
Propyl alcohol (iso)	VG	VG	VG	VG
Sodium hydroxide	VG	VG	VG	VG
Styrene	P	P	P	F
Stryene (100%)	P	P	P	F
Sulfuric acid	G	G	G	G
Tannic acid (65%)	VG	VG	VG	VG
Tetrahydrofuran	P	F	F	F
Toluene	F	P	P	F
Toluene diisocyanate	F	G	G	F
Trichloroethylene	F	F	P	G
Triethanolamine				
Tung oil	VG	P	F	VG
Turpentine	G	F	F	VG
Xylene	P	P	P	F

VG, very good; G, good; F, fair; P, poor.

C.1.12 Hearing protection

Worker exposure to too much noise usually depends on several factors, and establishing the need to provide hearing protection can be difficult. Hearing exposure depends on:

- The loudness of the noise as measured in decibels (dB).
- The duration of each employee's exposure to the noise.
- Whether employees move between work areas with different noise levels.
- Whether noise is generated from one or multiple sources.

Engineering controls and good work practices should lower worker noise exposure to acceptable levels, and if not, workers must wear appropriate hearing protection. Hearing protectors reduce only the amount of noise that gets through to the ears. Hearing protection devices must state their noise reduction rating (NRR). Workers exposed to occupational noise at or above 85 dB averaged over an eight-hour period are required to be in a hearing conservation program by their employer. Some types of hearing protection include the following:

- Single-use earplugs that are made of waxed cotton, foam, silicone rubber, or fiberglass wool.
- Pre-formed or molded earplugs which must be individually fitted by a professional and can be disposable or reusable.
- Earmuffs that require a perfect seal around the ear.

C.1.13 Proper use of PPE

PPE should be safely designed, constructed, and maintained clean. It should fit comfortably, encouraging worker use. It should fit the worker wearing the PPE properly. The collective efforts of both employers and employees will help in ensuring proper PPE use and maintaining a safe and healthy work environment. To ensure proper use of PPE, employers are required to train their workers on how to wear PPE correctly and care, clean, and maintain PPE. Employees must be trained to know at least the following:

- When PPE is required.
- What PPE is needed.
- How to properly put on, take off, adjust, and wear the PPE.
- The limitations of the PPE.
- Proper care, maintenance, useful life, and disposal of PPE.

Employers should make sure that each employee demonstrates an understanding of the PPE training as well as the ability to properly wear and use PPE before they are allowed to perform work requiring the use of the PPE.

PPE retraining is required in the following situations:

- If an employer believes that a previously trained employee is not demonstrating the proper understanding and skill level in the use of PPE.
- Changes in the workplace or in the type of required PPE that make prior training obsolete.

The employer must document the training of each employee required to wear or use PPE by preparing a certification containing the name of each employee trained,

Table C.1 PPE regulatory requirements

General industry (29 CFR 1910)	
1910 Subpart G - Occupational health and environmental control	1910.94, Ventilation 1910.95, Occupational noise exposure
1910 Subpart H - Hazardous materials	1910.120, Hazardous waste operations and emergency response
1910 Subpart I - Personal protective equipment	1910.132, General requirements
	1910.133, Eye and face protection
	1910.134, Respiratory protection
	1910.135, Head protection
	1910.136, Foot protection
	1910.137, Electrical protective equipment
	1910.138, Hand protection
	1910.140, Personal fall protection systems
1910 Subpart J - General environmental controls	1910.146, Permit-required confined spaces
1910 Subpart Q - Welding, cutting, and brazing	1910.252, General requirements
1910 Subpart Z - Toxic and hazardous substances	

the date of training, and a clear identification of the subject of the certification. Table C.1 contains a list of the regulatory requirements detailing PPE usage.

BIBLIOGRAPHY

1. Florida State University Department of Environmental Health and Safety. Retrieved from https://safety.fsu.edu/safety_manual/OSHA%20Glove%20 Selection%20Chart.pdf.
2. U.S. Department of Labor Occupational Safety and Health Administration. (2004). Personal Protective Equipment. (OSHA 3151-12R Retrieved from https://www.osha.gov/sites/default/files/publications/osha3151.pdf
3. Rosner, D., & Markowitz, G. (2020). A short history of occupational safety and health in the United States. *American Journal of Public Health, 110*(5), 622–628.
4. World Health Organization. (2020). Personal protective equipment.

Appendix D: Medical surveillance

D.I MEDICAL SURVEILLANCE REQUIREMENTS

The Occupational Safety and Health Act (OSHA) requires it to include provisions for medical examinations of employees in its standards. However, the precise test and examination criteria are not outlined in the OSHA. Instead, each standard has specific medical surveillance requirements. These are specific to the adverse health effects triggered by exposure to the hazardous substance. Medical surveillance is the systematic assessment of employees who are exposed to or expected to be exposed to occupational hazards. Employers undertake such medical surveillance over time for both their individual workers and groups of workers. The eventual goal of medical surveillance is to reduce occupational injury and illnesses of workers. According to OSHA medical screening and medical surveillance are two major strategies for optimizing employee health. Although the terms are often used interchangeably, they are quite distinct concepts. Medical screening is, in essence, only one component of a comprehensive medical surveillance program. The primary purpose of screening is early diagnosis and treatment of the individual and thus has a clinical focus. The primary purpose of surveillance is to detect and eliminate the underlying causes such as hazards or exposures of any discovered trends and thus has a prevention focus. Both can contribute significantly to the success of worksite health and safety programs. However, OSHA "medical surveillance" requirements are generally clinically focused (e.g., medical and work histories, physical assessment, biological testing) with information obtained from the clinical processes used in the monitoring and analysis elements of medical surveillance.

OSHA's policy regarding medical surveillance requirements is that the employer must make the medical examination available, at no cost, to the employee, but the employee is not required to take the examination. OSHA does not require an employer to force employees to take medical examinations. OSHA's regulations are intended as minimum standards. Employers can adopt more stringent requirements for themselves and, in doing so, may enforce mandatory participation in programs within the guidelines of labor/management relations.

Medical surveillance is conducted in conjunction with other employer measures, such as workplace practices and engineering controls, to prevent occupational injury and illness. Components of surveillance exams could include the following:

- OSHA respirator medical evaluation questionnaire
- Pulmonary function testing
- Respiratory fit testing
- Electrocardiogram (EKG)
- Lab work
- Exposure questionnaire
- Physical examination
- Fitness testing

In some cases, an employer may choose to provide no mandatory medical surveillance to monitor the health of workers when working with novel products and processes that can potentially expose them to hazardous products that may impact their health. The research and development industry is where the need for additional surveillance may be recommended. These situations may be rare but from time-to-time dose happens. In such case, the recommendation to provide surveillance is based on the recommendation of an occupational medicine (OM) physician.

Medical surveillance eligibility: OSHA recommends employers to offer medical surveillance to workers exposed to hazardous substances such as lead, asbestos, arsenic, bloodborne pathogens, radioactive materials insecticides that endanger worker health while at the workplace. Workers exposed to noise, chemical hazards, and toxic substances; requiring respiratory protection; and working in laboratories must also be given medical surveillance. Employees who work for employers engaged in hazardous waste operations and emergency response must also be provided with medical surveillance. A medical surveillance evaluation is conducted for such employees.

Table D.1 shows a list of hazardous substances included in the medical surveillance program.

Medical surveillance program: Employers are responsible for developing a comprehensive medical surveillance program to identify health hazards, develop health and safety programs, and increase the health and safety of employees. OSHA requires all costs of a medical surveillance program to be borne by the employer. Four key areas must be covered under this program.

a. **Surveillance:** Employers must set up a system to ensure employees undergo pre-employment screening before being allowed at a worksite with health and safety hazards. At the pre-employment screening, an individual's complete medical and illness history will be asked and recorded, a comprehensive physical examination will be conducted, and an evaluation will be done to understand the worker's ability to work while wearing personal protective equipment (PPE). Employers are required to provide medical surveillance when a worker resigns or is terminated from employment.

b. **Treatment:** A medical surveillance program must also cover treatment plans for workers for the different hazards that they may be exposed to. These treatment plans must consider preventive, emergency, and hospital care.

c. **Recording keeping:** Keeping detailed records of workers' medical examinations from pre-employment to the latest medical exam is an important part of the medical surveillance program

d. **Program review:** The effectiveness of a medical surveillance program can only be maintained when it is continually reviewed and revised or upgraded. Employers can use data gathered from worker injuries and health to enhance and upgrade the medical surveillance program annually.

The regulatory requirements for medical surveillance programs are prescriptive and therefore adherence to the requirements is important to provide comprehensive care when monitoring workers' health. Medical surveillance program requirements have been codified for chemicals, physical sources, and programs. Some examples of these program requirements are discussed below for a select few constituents.

D.2 COMPONENTS OF SOME MEDICAL SURVEILLANCE PROGRAMS

D.2.1 Asbestos

Per the OSHA standards for asbestos, exposure monitoring and medical surveillance of workers is required in the following situations:

- Workers are or will be exposed to airborne concentrations of fibers of asbestos at or above OSHA's exposure limits for a combined total of 30 or more days per year.
- Workers perform work that disturbs asbestos-containing material (ACM) or presumed asbestos-containing material (PACM) for a combined total of 30 or more days per year.
- For workers that wear negative-pressure respirators – the employer must provide a medical evaluation in accordance with Appendix D of OSHA's asbestos standards.

An asbestos physical consists of the following:

- Comprehensive medical exam
- Work and medical history review
- Chest X-ray with B Reader

- Pulmonary function test (PFT) with blood pressure
- OSHA Respirator Medical Questionnaire
- Any laboratory test that the examining physician deems necessary upon review of medical exam results and medical work history

D.2.2 Benzene

- Medical surveillance program is designed to observe adverse health effects of benzene exposure.
- The main effects of benzene exposure manifesting clinically are pancytopenia, aplastic anemia, and leukemia. A detailed occupational history which includes a history of past work exposure to benzene or any other hematological toxin, a family history of blood dyscrasias including hematological neoplasms, a history of blood dyscrasias, bleeding abnormalities, and abnormal functions of formed blood elements, a history of renal or liver dysfunction, a history of medications routinely taken, a history of previous exposure to ionizing radiation, and a history of exposure to marrow toxins outside of the employee's current work situation. The employee must provide to the examining physician as complete an occupational history as possible for the period prior to the current employment.
- Symptoms and signs of benzene exposure can be non-specific and so a detailed history and appropriate investigative procedures will enable a physician to rule out or confirm conditions that place the employee at increased risk. To assist the examining physician regarding which laboratory tests are necessary and when to refer an employee to the specialist, OSHA has established hematological guidelines to be followed.
- Initial examinations are to be provided to employees at the time of initial assignment before they are permitted to enter or continue working in a workplace in which they may be exposed to benzene and periodic examinations annually thereafter.

D.2.3 Blood-borne pathogens

- Bloodborne pathogens are infectious microorganisms in human blood that can cause disease in workers. These pathogens include, but are not limited to, hepatitis B virus (HBV), hepatitis C (HCV), and human immunodeficiency virus (HIV). Needlesticks and other sharp-related injuries may expose workers to bloodborne pathogens.
- In accordance with the Health Insurance Portability and Accountability Act or HIPAA, effective April 14, 2003, all patient-related medical information will be kept confidential.
- Employers must enroll employees with known or potential exposure to blood or human blood products in Bloodborne Pathogens Program

within 10 days of hire. This includes all employees anticipated to encounter bodily fluids. This program also provides Hepatitis B vaccinations to employees.

D.2.4 Lead

- The employer must institute a medical surveillance program for all employees who are or may be exposed at or above the action level for more than 30 days in any 12 consecutive months.
- Initial medical surveillance is used to check the amount of lead in an employee's bloodstream. This is referred to as biological monitoring. The two blood tests used in biological monitoring are the blood lead level test and the zinc protoporphyrin (ZPP) test. Initial medical surveillance shall be provided at no cost to employees involved in lead tasks or if an employee is exposed to lead on the job any one day at or above the action level.
- Employees shall be placed in an ongoing medical surveillance program if it is anticipated that the employee will be exposed to lead on the job at or above the action level for more than 30 days in any continuous 12-month period.
  - Interview about the employee's work and medical history
  - Complete physical exam
  - Blood pressure check
  - Blood tests that will show blood level, ZPP, hemoglobin, and hematocrit (anemia test)
  - Blood urea nitrogen and serum creatinine (kidney test)
  - Routine urinalysis (kidney and protein check)
  - Any additional test that the doctor needs to do to determine how lead has or could affect the employee
- The employer must maintain an accurate record for each employee subject to medical surveillance including the following:
  - A description of the employee's duties
  - A copy of the physician's written opinions
  - The results provided by the examining physician of any airborne exposure monitoring done for the representative employee and all others represented; and any employee medical complaints related to lead exposure
- The employer must also obtain and furnish to the employee the physician's written opinion:
  - Whether the employee has any detected medical condition that would place his or her health at increased risk from lead exposure
  - Any special protective measures or limitations on worker's exposure to lead
  - Any limitation on respirator use
  - Results of blood lead determination

D.2.5 Noise

- Noise medical surveillance is a surveillance that focuses on hearing only. Exposure to loud noise at work can cause deterioration in hearing over time known as noise-induced hearing loss (NIHL).
- The purpose of a hearing surveillance program is to detect early signs of NIHL to allow intervention before significant hearing loss occurs. The program also identified individuals at risk of hearing loss and is a measure of the effectiveness of noise control measures.
- Workers whose noise exposures equal or exceed an 8-hour time-weighted average (TWA) of 85 decibels are enrolled in a *Hearing Conservation Program*.
- Noise medical surveillance forms part of the employer's overall noise control program with the aim of reducing exposure to loud noise using engineering, administrative, educational and use of hearing protection equipment (HPE).
- A noise medical surveillance program involves periodically and systematically assessing the hearing of all employees who may be exposed to noise.
  - The assessment includes a questionnaire looking for symptoms of NIHL and other hearing issues.
  - A clinical examination of the ear canal and an audiogram is done.
  - Results are analyzed for each employee and for work areas.

Repeated testing over time will show patterns of concern for individual employees and specific work areas; this allows appropriate individual and group intervention to protect hearing.

D.2.6 Respiratory protection

- Medical surveillance and screening may be recommended or required based on job duties that require respiratory protection or mandatory respirator use.
- The purpose of medical surveillance is to protect workers who might have to use respirators. A complete physical examination of each respirator user is not usually required, but an initial medical examination and an annual review of medical status are required.
- If any change in work conditions or operational procedures are planned that would affect an individual's exposure to hazardous airborne contaminants (e.g., changes in hazardous substances, new equipment, etc.), the employer will evaluate the use of respiratory protection under the new circumstances.
- Medical surveillance is required for employees who wear a respirator for 30 days or more a year.
- Employees who are required to wear a respirator will be enrolled in the Respiratory Protection Program. The program elements include the following:

- Training
- Medical evaluation
- Respirator fit testing
- Program evaluation
- Recordkeeping

D.2.7 HAZWOPER

- The main purpose of the Hazardous Waste Operations and Emergency Response (HAZWOPER) medical surveillance program is to make sure employees are fit to work in hazardous waste operations. It ascertains an employee's health status when first assigned to hazardous waste tasks so that healthcare provider can continuously monitor employee's health.
- The HAZWOPER Standard applies to employers whose employees respond, investigate, or clean up hazardous materials.
- Employers are responsible for developing a comprehensive medical surveillance program to identify any health hazards and develop health and safety programs for their employees.
- A medical surveillance program is required for
  - Employees who are or may be exposed to hazardous substances or health hazards at or above the permissible exposure limits (PELs) as set by OSHA
  - Employees who are injured, become ill, or develop signs or symptoms due to possible overexposure involving hazardous substances or health hazards from an emergency response or hazardous waste operation
  - Employees on the hazmat team

D.3 OCCUPATIONAL MEDICINE

OM focuses on the interface of the workplace and health. Occupational medicine doctors (OMD) are skilled at combining individual patient care with prevention. These physicians generally take a population-based health approach and are usually engaged in all aspects of workers' health and the workplace. They are actively engaged in occupational health issues that involve chemical, biological, physical, and psychosocial hazards. The specialty encompasses workers' wellness, disease prevention, and occupational injury and illness care.[1]

Realizing the importance of OM, many companies have on location a clinic for employees that is staffed with at least one occupational physician. If not on site, they have available to their employees' OM physicians through a contract with a health clinic or with a private practice physician's office. These physicians work closely with safety and health professionals in their plight to facilitate the safety and health of workers.

D.4 MEDIAL SURVEILLANCE IMPLEMENTATION

A comprehensive medical surveillance program can be administered directly by the employer. This is where the employer has an onsite clinic where the physician and medical staff are employed by the company. Many large employers choose this path because it offers flexibility in affording the employees to maintain their surveillance requirements with less time away from the worksite. Some of the benefits of the in-house options are as follows:

- Employees spend least time away from work for medical surveillance medical appointments.
- Medical providers have some understanding of the business and can recognize exposure's potential catalyst in the event of a work-related injury or illness. This knowledge can facilitate and expedite proper treatment.
- The ability to quickly respond to work-related injuries is needed.

Potential downfall of this method may include the following:

- Perception of some employees that some medical decisions may be made to benefit the employer's desire to keep productivity moving
- Company may be responsible for medical insurance for physicians. If not paid by the company, physicians will be responsible to obtain medical insurance if deem necessary
- Some employees may miss appointments because they may not see the importance of keeping appointments as they can just reschedule without penalty (company being charged or missed appointments)

Contracting with a medical provider is another option that is often taken by small companies and large companies. This option seems to be the most used option because it may be perceived by some employers the least expensive and litigious of the two. With this option the company has a contract with a primary provider who administers the medical surveillance program and maintains surveillance records as would any other medical records that is retained for other patients. Some providers take advantage of the mobile medical service. This service is customized based on the needs of the employer and has scheduled time to bring the medical surveillance program option to the workplace.

Some of the benefits of the contract-out options are as follows:

- Company has no need to secure medical malpractice insurance for internal physicians.
- The expense of staffing and maintaining medical equipment and supplies will not be necessary.
- The employee may believe that the independence between medical providers and employer is beneficial to the level of services provided.

Potential downfall of this method may be the following:

- The increased amount of time an employee may spend away from work to attend medical appoints.
- Employees may take extra time for appointment because management has no way of knowing how much time an appointment will take.
- The company may be billed for appointments that are missed by employees.
- Employee may not have quick access to medical records.

D.5 THE ENROLLMENT PROCESS

Before enrollment can begin, a knowledge of the hazard profile for a process or a particular work group is required. The hazard profile should provide an indication of potential exposures that a person or work group may be subjected. Once it has been determined the workers who should be enrolled in a medical surveillance program, it is the supervisor's responsibility to start the enrollment process. They assist supervisors in providing the needed information to the medical provider, completion of a medical surveillance form should be completed for each enrollee. An example of a medical surveillance form is included at Figure D.1. The medical surveillance form should be developed with the specific company in mind including the substances or activities that specifically applies to the constituents needed to be subject to a medical surveillance program as outlined by OSHA. Additional information will be collected by the medical provider because the type of additional needed is most likely governed by the HIPPA regulations protecting confidentiality of workers health information.

Medical Surveillance Enrollment

Name: _____   Date: _____

Department: _____   Work Group: _____

Description of work activity: (include time performing activity)

_____

_____

_____

_____

Engineering controls: _____

PPE: _____

Has an exposure potential assessment been form: Yes   No

Chemical of Concern: (circle all that apply)

Acrylonitrite	Arsenic	Asbestos	Benzene
Bloodborne Pathogens	1,3-Butadiene	Cadmium	Chromium VI
Formaldehyde	HAZWOPER	Lead	Methylene Chloride
Noise	Respiratory Protection	Vinyl Chloride	

Other: _____

Completed By: _____   Date: _____

Supervisor Signature: _____   Date: _____

Figure D.1 Example of medial surveillance enrollment form.

Chemical	Pre-placement exam	Periodic exam	Regulation
Acrylonitrile	Yes	Yes – annual	1910.1045(n); 1926.1145; 1915.1045
Arsenic (inorganic)	Yes	Yes – annual	1910.1018(n); 1926.1118; 1915.1018*
Asbestos (general industry)	Yes	Yes – annual	1910.1001(l)
Asbestos (construction and shipyards)	Yes	Yes – annual (more frequently if determined to be necessary by a physician)	1926.1101(m); 1915.1001
Benzene	Yes	Yes – annual	1910.1028(i); 1926.1128; 1915.1028*
Bloodborne pathogens	No – Hepatitis B (HBV) vaccine to be offered	No	1910.1030(f)
1,3-Butadiene	Yes	Yes	1910.1051(k); 1926.1151*
Cadmium	Yes	Yes	1910.1027(l); 1926.1127; 1915.1027; 1928.1027*
Carcinogen (suspect)	Yes	Yes – annual	1910.10031016(g); 1926.1103; 1915.10031016*
Chromium (VI), hexavalent chromium	Yes	Yes	1910.1026(k); 1926.1126(i); 1915.1026(i)
Coke oven emissions	Yes	Yes	1910.1029(j)
Compressed air environments	Yes	Yes	1926.803(b)
Cotton dust	Physical exam not specified	Physical exam not specified	1910.1043(h)
1,2-Dibromo3chloropropane	Yes	Yes	1910.1044(m); 1926.1144; 1915.1044
Ethylene oxide	Yes	Yes – annual	1910.1047(i); 1926.1147
Formaldehyde	Yes	Yes	1910.1048(l); 1926.1148; 1915.1048
HAZWOPER	Yes	Yes – annually or at the discretion of a physician's	1910.120(f); 1926.65

(Continued)

(*Continued*)

Chemical	Pre-placement exam	Periodic exam	Regulation
Hazardous chemicals in laboratories	When required by other standards	When required by other standards	1910.1450(g)
Lead	Yes – except in construction industries; construction requires initial blood tests only	Yes	1910.1025(j); 1926.62
Methylene chloride	Yes	Yes	1910.1052(j); 1926.1152
Methylenedianiline	Yes	Yes – Annual	1910.1050(m)
Noise	No – baseline audiogram required within six months of exposure at or above 85dB	Annual audiogram testing required	1910.95(g); 1926.52
Respiratory protection	Evaluation questionnaire or exam; follow up exam when required	Yes – in specific situations	1910.134(e); 1926.103
Vinyl chloride	Yes	Yes	1910.1017(k); 1926.1117

BIBLIOGRAPHY

1. Baker, B., Kesler, D., & Guidotti, T. (2020). Occupational and environmental medicine: Public health and medicine in the workplace. *American Journal of Public Health*, 110(5), 636–637. https://doi./10.2105/AJPH.2020.305625.
2. U.S. Department of Labor Occupational Safety and Health Administration. (2014). Medical Screening and Surveillance Requirements in OSHA Standards: A Guide (OSHA 3162-01R) Retrieved from https://www.osha.gov/sites/default/files/publications/osha3162.pdf

Index

Note: **Bold** page numbers refer to tables; *italic* page numbers refer to figures.

Printed in the United States
by Baker & Taylor Publisher Services